Data Fusion and Data Mining for Power System Monitoring

Data Fusion and Data Mining for Power System Monitoring

Arturo Román Messina

CRC Press
Taylor & Francis Group
Boca Raton   London   New York

CRC Press is an imprint of the
Taylor & Francis Group, an **informa** business

First edition published 2020
by CRC Press
6000 Broken Sound Parkway NW, Suite 300, Boca Raton, FL 33487-2742

and by CRC Press
2 Park Square, Milton Park, Abingdon, Oxon, OX14 4RN

ISBN: 9780367333676 (hbk)
ISBN: 9780429319440 (ebk)

Typeset in Palatino
by Lumina Datamatics Limited

Contents

Section II   Advanced Projection-Based Data Mining and Data Fusion Techniques

Section III   Challenges and Opportunities in the Application of Data Mining and Data Fusion Techniques

Preface

Data mining and data fusion have become the focus of much research in recent years, especially in areas of applications for which high-dimensional data sets are collected and analyzed, such as in the case of wide-area measurement and monitoring systems. The interest in these methods arises because of their ability to process, analyze, and classify large amounts of heterogeneous high-dimensional data.

With advances in communications technology and computer power, the last few decades have witnessed the emergence of a new generation of wide-area monitoring systems with the ability to monitor system state in near real-time, assess power system health, and aid in decision making.

Data mining is increasingly being used to extract and classify dynamic patterns from multisensor, multiscale data and may be used as a preprocessing step for other applications. Fusing multimodal data, on the other hand, allows analysts to combine data from different sensors to provide a robust and more complete description of a dynamic process or phenomena of interest.

This book deals with the development and application of advanced data mining and data fusion techniques for monitoring power system oscillatory behavior. Using a common analytical framework, this book examines the application of data mining and data fusion techniques to monitor key system states and assess system health. Examples of applications to a wide variety of simulated and measured data are provided. Many results have general applicability to other processes or physical phenomena.

This book is organized into three sections. Section I, encompassing Chapters 1 and 2, gives an overview of modern wide-area monitoring systems and examines mining and fusion architectures. The general concepts underlying the notion of data mining and data fusion are also introduced.

In Section II, attention is focused on the modeling of spatiotemporal data and the development of advanced projection-based data mining and data fusion techniques. Chapters 3 through 7 examine the problem of spatiotemporal modeling, stressing the use of methods for nonlinear dimensionality reduction, clustering, and feature extraction. Typical analysis techniques are reviewed, and methods to mine, fuse, and analyze data are examined using a rigorous analytical framework. A variety of examples are described to illustrate the application of these methods.

Section III discusses challenges and opportunities in the application of data mining and data fusion techniques to transmission and distribution data. Chapter 8 examines the use of visualization techniques to monitor system oscillatory behavior. Chapter 9 discusses challenges and opportunities for research, with emphasis on application to power system monitoring.

Finally, Chapter 10 examines the application of various data mining and data fusion techniques to monitor power system dynamic behavior in realistic, large-scale test power systems.

Primarily, the book is intended for advanced undergraduate and graduate courses, as well as for researchers, utility engineers, and advanced teaching in the fields of power engineering, signal processing, and the emerging field of spatiotemporal analysis and modeling.

Arturo Román Messina

Author

Arturo Román Messina earned his PhD from Imperial College, London, UK, in 1991. Since 1997, he has been a professor in the Center for Research and Advanced Studies, Guadalajara, Mexico. He is on the editorial and advisory boards of Electric Power Systems Research, and Electric Power Components and Systems. From 2011 to 2018 he was Editor of the *IEEE Transactions on Power Systems* and Chair of the Power System Stability Control Subcommittee of the Power Systems Dynamic Committee of IEEE (2015–2018). A Fellow of the IEEE, he is the editor of *Inter-Area Oscillations in Power Systems – A Nonlinear and Non-stationary Perspective* (Springer, 2009) and the author of *Robust Stability and Performance Analysis of Large-Scale Power Systems with Parametric Uncertainty: A Structured Singular Value Approach* (Nova Science Publishers, 2009) and *Wide-Area Monitoring of Interconnected Power Systems* (IET, 2015).

List of Abbreviations

BSI	blind source identification
BSS	blind source separation
CCA	canonical correlation analysis
DHR	dynamic harmonic regression
DM	diffusion maps
DMD	dynamic mode decomposition
EOF	empirical orthogonal function
ERP	event-related potential
HHT	Hilbert–Huang Transform
ICA	independent component analysis
IVA	independent vector analysis
JIVE	joint and individual variation explained
KF	Kalman filter
MB	multiblock
PCA	principal component analysis
PDC	point data collector
PLS	partial least squares
PMU	phasor measurement unit
POD	proper orthogonal decomposition
SOBI	second-order blind source separations
SVC	static VAr compensator
SVD	singular value decomposition
SVM	support vector machine
WAMS	wide-area measurement systems

Section I

Overview of Modern Wide-Area Monitoring Systems

1

Introduction

1.1 Introduction to Power System Monitoring

The past few decades have witnessed important advances in the development of wide-area measurement and monitoring systems that enable better operational awareness of the real-time condition of the grid and have the potential to improve system reliability (Bobba et al. 2012; Hauer et al. 2007; Phadke and Thorp 2008; Zuo et al. 2008; Atanackovic et al. 2008). At the heart of modern WAMS are advanced analytical techniques with the ability to reduce information complexity, detect trends, identify and locate disturbances, and control impending threats (Begovic and Messina 2010).

A significant element of this major thrust is the development of multichannel sensors deployed throughout the system to monitor system status; examples include Global Position System (GPS) based time-synchronized phasor measurement units (PMUs), frequency disturbance recorders (FDRs) and their associated wide-area frequency measurement networks (Zhang et al. 2010; Burgett et al. 2010; Chai et al. 2016), micro-PMUs, protection relays, fault recorders, and Supervisory Control and Data Acquisition (SCADA) systems.

Synchronized, high-resolution distributed monitoring and measurement systems improve system observability and can be used to observe, assess, and mitigate system threats (Karlsson et al. 2004). Simultaneous analysis of large volumes of high-dimensional data coming from different sources is challenging, since many power system dynamic processes evolve in space and time with complicated structures. Much like the case of observational data in other disciplines (such as, neuroscience, climate and environmental science, health care, and seismology), the resulting data often exhibit non-stationary dependence structures and show varying properties in different spatial regions and time periods (Atluri et al. 2018).

The huge size of modern power systems (combined with the presence of resonances, control actions, and uncertain load behavior) can make the application of conventional data mining/data fusion techniques ineffective or lead to poor accuracy and interpretability of results. Wide area dynamic recorders have different measurement characteristics, resolution, and data coverage, which makes the joint analysis of measured data necessary.

Extracting relevant system dynamics from these vast arrays of data is a difficult problem, which requires the integration of sophisticated communications, advanced processing capabilities, and vast data storage technologies (Allen et al. 2013). The advent of new sensor devices and advances in data collection and storage technologies prompt the need for the development of efficient data-driven decision-making algorithms. Furthermore, local information collected by PMUs or other dynamic recorders may capture different scales or varying degrees of nonlinear behavior (or non-stationarity) and they may exhibit different levels of missing data (NERC 2010).

Data mining and data fusion are two active areas of research with application to many fields (Thomopoulos 1994). This chapter provides a brief introduction to the problem of wide-area monitoring. Features relevant for wide-area oscillation monitoring are reviewed, and key technological advances are discussed in terms of how they can be used to monitor power system status.

1.2 Wide-Area Power System Monitoring

Over the last few decades, various forms of wide area monitoring systems have been developed to monitor system behavior, including wide-area frequency measurement networks, PMU-based WAMS, and SCADA systems. Relevant experience with the application of wide-area frequency measurements is described in (Zhang et al. 2012).

In its most elementary form, a wide-area monitoring system is an intelligent, continuous identification overseer of power system status (Messina 2015). Current WAMS architectures are evolving from advanced monitoring to more intelligent, wide-area control systems with the capacity to forecast system behavior, identify threats, and develop corrective actions (Hauer et al. 2007). Opportunities and challenges in the integration of synchrophasor technology to energy management systems (EMS) at the system control center are discussed in Elizondo et al. (2012) and Atanackovic et al. (2008).

Conceptually, a WAMS consists of different components such as dynamic recorders, advanced communication systems, and signal processing techniques. Advances in sensor technologies and data acquisition platforms have led to massive amounts of data that needs to be integrated for efficient and accurate power system monitoring. This data provides new opportunities to monitor and assess power system health and integrity but also opens new challenges for analysis that must be addressed.

A schematic depiction of the functional elements of a WAMS is shown in Figure 1.1. The typical WAMS structure is hierarchical and centralized and can be divided into two major levels or layers: regional and global. At a local level, measured information is automatically collected, synchronized, and archived by a monitoring and control center known as Phasor Data

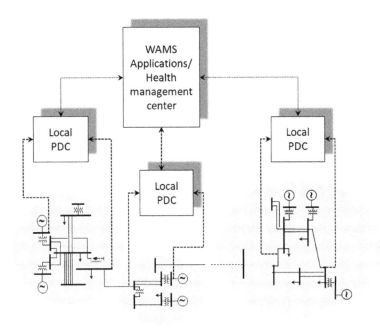

FIGURE 1.1
Illustration of a wide-area measurement system.

Concentrator (PDC) (Phadke and Thorp 2008; Kincic et al. 2012). This information is then sent to a global data concentrator for real-time dynamics monitoring, wide-area control, and wide-area protection. In centralized architectures, the super PDC provides a platform for different WAMS applications (Zuo et al. 2008).

Successful implementation of real-time monitoring schemes based on synchrophasor technology demands the integration of several levels of triggering and settings that detect deteriorating system conditions in the presence of normal system behavior.

1.2.1 WAMS Architectures

In modern WAMS architectures, system health is continuously monitored and assessed to identify threats and characterize associated risks (prognosis). Visualization and awareness systems are also of vital importance in many applications (Pan et al. 2015).

Various conceptualizations of the monitoring process exist in the literature and different applications may require fusion at different levels. Figure 1.2 shows a generic three-layer WAM structure inspired by Hauer et al. (2007) and Messina (2015).

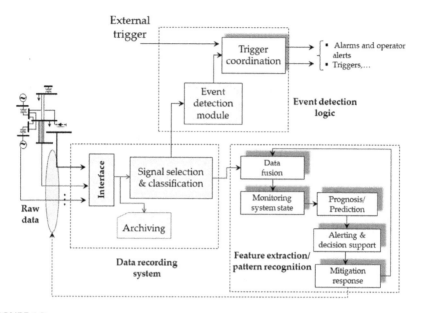

FIGURE 1.2

Generic WAMS structure. (Based on Hauer, J. et al., A perspective on WAMS analysis tools for tracking of oscillatory dynamics, *2007 IEEE Power Engineering Society, General Meeting*, Tampa, FL, June 2007; Messina, A., *Wide-Area Monitoring of Interconnected Power Systems*, IET Power and Energy Series 77, London, UK, 2015.)

The model includes the following three (interrelated) tasks: (1) data acquisition and data management module, (2) event logic detection and signal processing, (3) health monitoring and assessment. The reader is referred to Hauer et al. (2007) and Messina (2015) for a more detailed treatment of the first two issues. Attention here is focused on health monitoring and assessment.

Health monitoring and assessment implies the integration of methodologies to extract indications or assessments of critical system characteristics directly from measured data. It usually integrates three main objectives: measurement, assessment, and decision making. At every level of this structure, there will be the need for the following basic processes to exist:

- Dimensionality reduction algorithms,
- Association analysis,
- Clustering and classification,
- Anomaly detection,
- Regression and prediction and forecasting, and
- Decision support.

1.2.2 Synchronized Sensor Technology

Sensors are a critical and complex component of modern WAMS. Sensors arrays can be either local or distributed (i.e., using remote connections to the monitoring system) (Messina 2015).

The latest generation of PMUs and other power system dynamic recorders can measure multiple parameters, such as voltage, frequency, phase and active/reactive power, and frequency and voltage phase angle (in the case of frequency recorders) at varying temporal resolutions. The reporting rate of current PMUs is up to 60 and 120 phasors per second or 1 phasor every 16.67 (0.0083) milliseconds (Lee and Centeno 2018). This shows that current PMUs provide the ability to observe slow system behavior.

1.2.3 Integrated Communication Across the Grid

Communication infrastructure and signal processing techniques are critical enablers of modern WAMS. Accurate transmission of data requires modern WAMS architectures to include advanced high-speed communication systems and protocols, along with advanced data analysis techniques. Determining suitable criteria for designing the communications systems is an important issue that has received limited attention (Zhang et al. 2012).

In parallel to this trend, distributed data mining and data fusion techniques are needed to process highly distributed data (Thomopoulos 1994; Phadke and Thorp 2010). Even with simplifications and approximations of various types, dimensionality reduction techniques are needed to analyze, correlate, and synthesize large volumes of high dimensional data.

1.3 Data Fusion and Data Mining for Power Health Monitoring

Integrating complex, heterogeneous data from multiple multichannel sensors is a difficult problem due to both conceptual and computational issues. PMU data are heterogeneous and may exhibit differing spatial and temporal scales that make global analysis difficult. In many applications, heterogeneous or multisource data acquired through sensor arrays, such as WAMS or other sensors, are analyzed separately. Fusing information from different sources, however, may provide additional insight into various aspects of system dynamic behavior and lead to improved characterization of data relationships.

Each type or modality of data (i.e., phase, voltage magnitudes and phase angles, frequency, and active/reactive power) may provide unique information that is complementary to other data types. In addition, joint analysis of multitype data allows the analysis of interactions or association between data sets of different modalities (Dutta and Overbye 2014; Fusco et al. 2017). Even if the data is from the same modality, such as speed deviations in wind farms and synchronous generation, combined analysis of these data types can provide new insights into system behavior and enable the analysis of correlation measures (Yang et al. 2014).

Data mining and data fusion techniques are often used to extract dynamic patterns from data. Although data mining and data fusion techniques may provide efficient solutions to analyze large heterogeneous data sets, there are a number of challenges and limitations that need to be addressed:

- Sensor data, especially in the case of wide-area measurements, is often correlated. As a result, some features may be nearly irrelevant or redundant, which may result in numerical problems in the application of some methods.

- Data mining and data fusion techniques may increase latency and impact the burden on communication infrastructure (Khaleghi et al. 2013; Hall and Steinberg 2001).

- The performance of the methods may be affected by noise, uncertainty, outliers or spurious (missing) data and other artifacts. Techniques are needed to preprocess and clean data to improve the efficiency of the methods. As discussed by Hall and Steinberg (2011), there are no perfect algorithms that are optimal under all circumstances.

- Although progress has been made in recent years, near real-time implementation of data mining and data fusion techniques remains a challenging task.

- Most data fusion architectures are centralized or hierarchical in nature. With the advent of distributed renewable generation and more complex heterogeneous sensor networks, there has been a surge of interest in strategies to mine and fuse data to monitor global system behavior (Yang et al. 2014). Also, distributed algorithms are needed for optimal sensor placement.

- Current monitoring and analysis techniques may not be appropriate to handle the much-increased dimensionality of sensor data and new developments are needed both in visualization tools and in spatial/temporal analysis tools.

In summary, data fusion and data mining technologies have a key role to play in the development of future WAMS and Smart Grids (Zhang et al. 2010; Bhatt 2011; Hou et al. 2019). The effective integration of data is essential to improving power system health.

1.4 Dimensionality Reduction

Wide-area PMU data are characterized by many measured variables and the record lengths are growing rapidly. Although the number of sensors and associated variables is large, the collected data set is often highly redundant and often only a few underlying variables or patterns are relevant to global behavior.

As discussed in (Sacha et al. 2017), dimensionality reduction is a core building block in visualizing and classifying high-dimensional data. Dimensionality reduction facilitates both data classification and data communication, and it also reduces measurement and storage requirements (Hinton and Salakhutdinov 2006); this enables researchers to find features that are not apparent in the original unreduced space as well as to select good observables for power system monitoring (Sacha et al. 2017).

In the literature, two main approaches for reducing system dimensionality and selecting key system variables are selection-based and projection-based methods (Cui et al. 2013). These issues will be discussed in more detail in subsequent chapters.

1.5 Distribution System Monitoring

In recent years there has been rapid development and increasing use of technologies for monitoring distribution system networks and microgrids. Along with the increase of information from smart meters, other sensors have been developed for analysis of distribution data such as micro-phasor measurement units. These devices provide synchronized, accurate measurements of voltage (current) magnitudes and phase angles (Von Meier et al. 2017). Distribution PMUs are expected to coexist with other measurement devices such as smart meters and SCADA, which makes the application of data mining and data fusion techniques appealing. Analyzing and monitoring distribution data requires data mining technologies designed for distributed applications.

Potential applications include post-mortem event detection and classification (Pinte et al. 2015; Von Meier et al. 2017), topology and cyber-attack detection (Von Meier et al. 2017), improvement in network model accuracy (Roberts et al. 2016), characterization of distributed generation, and phasor-based control, among other uses. Other emerging applications include monitoring, protection, and control of distribution networks (Roberts et al. 2016). Also, as discussed by Jamei et al. (2018), proper placement of PMU sensors plays a key role in their ability to detect anomalous events.

1.6 Power System Data

At the core of a wide-area monitoring system is data. Wide-area monitoring systems provide high quality, multisource, and multitype spatiotemporal data. In many applications, data from sensor arrays can be also complemented with other data, such as historical information, climate information, and the status of system breakers.

Power system data, however, can bring forth unique challenges and opportunities for power system monitoring. First, data records are becoming increasingly long and may exhibit missing points, trends, and other artifacts that make the application of conventional analysis techniques difficult. In addition, typical observational data are correlated with each other and often exhibit some degree of dependency, especially in the case of multichannel data recorded at nearby locations. This essentially means that the intrinsic dimensionality of the input space (data space) is much less than the number of sensors or measurements. Modern PMUs can provide multichannel data (e.g., data containing several physical parameters). This has several implications in some data mining/data fusion techniques that assume independence among observations (Atluri et al. 2018).

Second, the increasing availability of sensors with different temporal (and spatial) resolution requires the development of advanced data fusion and data mining techniques with the ability to interpret the collected data which results in improved dynamic characterization of system behavior. A related problem is that of determining good observables in high-dimensional data. For wide-area applications, reduction methods can be used to detect good globally reduced coordinates that best capture the data variability. In this regard, it is crucial to identify anomalous or abnormal events, as well as the extent and distribution of the ensuing events.

Finally, with the advent of distributed renewable generation and more complex wide-area monitoring systems, there has been a surge of interest in utilizing hierarchical, structured, distributed monitoring systems for assessing global system health (Zhang et al., 2007, Stewart et al. 2017). Data from different sources can be used to check consistency, provide context, and improve understanding of the underlying causes associated with an observed phenomenon. But, as noted in Section 1.5, distributed data sent to regional PDCs may affect communication bandwidths, introduce further latency, and introduce new temporal and spatial scales that compound monitoring issues.

Data complexity arises from several factors including its diversity, heterogeneity, and nature. These factors render the collected data sets to be uncertain and imprecise. The ever-growing number of PMUs and other dynamic recorders or sensors, on the other hand, is creating vast amount of data that must be correlated and processed. Multisensor data may capture complementary aspects of the observed phenomenon and may enable one to extract a more reliable and detailed description of the measured phenomenon (Katz et al. 2019).

The implementation of such systems requires a combination of sensor data fusion, feature extraction, classification, and prediction algorithms. In addition, new system architectures are being developed to facilitate the reduction of wide bandwidth sensor data resulting in concise predictions of the ability of the system to complete its current or future missions.

1.7 Sensor Placement for System Monitoring

Effective assessment and monitoring of the dynamic performance of the power system requires wide-area information from properly discriminated sensor arrays. Sensors should be located appropriately to achieve maximum detection performance and allow efficient collaboration among sensors.

Sensor placement involves the solution of two main problems: (1) identifying the best signals to observe system behavior and (2) determining the number and location of sensors.

A related issue is that of determining redundant measurements. While some authors regard redundant information as a means of strengthening or confirmation, redundant information is also seen as a repetition or redundancy that can be ignored.

References

Allen, A., Santoso, S., Muljadi, E., Algorithm for screening phasor measurement unit data for power system events and categories and common characteristics for events seen in phasor measurement unit relative phase-angle differences and frequency signals, Technical Report, NREL/TP-5500-58611, National Renewable Energy Laboratory, Golden, CO, August 2013.

Atanackovic, D., Clapauch, J. H., Dwernychuck, G., Gurney, J., First steps to wide-area control: Implementation of synchronized phasors in control center real-time applications, *IEEE Power & Energy Magazine*, 61–68, January/February 2008.

Atluri, G., Karpatne, A., Kumar, V., Spatio-temporal data mining: A survey of problems and methods, *ACM Computing Surveys (CSUR)*, 51(4), 1–41, 2018.

Begovic, M. M., Messina, A. R., Wide-area monitoring, protection and control, *IET Generation, Transmission & Distribution*, 4(10), 1083–1085, 2010.

Bhatt, N. B., Role of synchrophasor technology in the development of a smarter transmission grid, *2011 IEEE Power and Energy Society General Meeting*, Minneapolis, MN, 2011.

Bobba, R. B., Daggle, J., Heine, E., Khurana, H., Sanders, W. H., Sauer, P. Yardley, T., Enhancing grid measurements, *IEEE Power & Energy Magazine*, 67, 73, January/February 2012.

Burgett, J., Conners, R. W., Liu, Y., Wide-area frequency monitoring network (FNET) architecture and applications, *IEEE Transactions on Smart Grid*, 1(2), 159–167, 2010.

Chai, J., Liu, Y., Guo, J., Wu, L., Zhou, D., Yao, W., Liu, Y., King, T., Gracia, J. T., Patel, M., Wide-area measurement data analytics using FNET/GridEye: A review, *2016 Power Systems Computation Conference*, Genoa, Italy, June 2016.

Cui, M., Prasad, M., Li, W., Bruce, L. M., Locality preserving genetic algorithms for spatial-spectral hyperspectral image classification, *IEEE Journal of Elected Topics in Applied Earth Observations and Remote Sensing*, 6(3), 1688–1697, June 2013.

Dutta, S., Overbye, T. J., Feature extraction and visualization of power system transient stability results, *IEEE Transactions on Power Systems*, 29(2), 966–973, 2014.

Elizondo, D., Gardner, R. M., León, R., Synchrophasor technology: The boom of investments and information flow from North America to Latin America, *2012 IEEE Power Engineering Society General Meeting*, San Diego, CA, July 2012.

Fusco, F., Tirupathi, S., Gormally, R., Power systems data fusion based on belief propagation, *2017 IEEE PES Innovative Smart Grid Technologies Conference Europe (ISGT-Europe)*, Torino, Italy, September 2017.

Hall, D. L., Steinberg, A., Dirty secrets in multisensory data fusion, Defense Technical Information Center, Technical report, https://apps.dtic.mil/dtic/tr/fulltext/u2/a392879.pdf, 2001.

Hauer, J., Trudnowski, D. J., DeSteese, J. G., A perspective on WAMS analysis tools for tracking of oscillatory dynamics, *2007 IEEE Power Engineering Society, General Meeting*, Tampa, FL, June 2007.

Hinton, G. E., Salakhutdinov, R. R., Reducing the dimensionality of data with neural networks, *Science*, 313, 504–507, July 2006.

Hou, W., Zhaolong, N., Guo, L., Zhang, X., Temporal, functional and spatial big data computing framework for large-scale smart grid, *IEEE Transactions on Emerging Topics in Computing*, 7(3), 369–379, 2019.

Jamei, M., Scaglione, A., Roberts, C., Stewart, E., Peisert, S., McParland, C., McEachern, A., Anomaly detection using optimally placed μPMU sensors in distribution networks, *IEEE Transactions on Power Systems*, 33(4), 311–3623, 2018.

Karlsson, D., Hemmingsson, M., Lindahl, S., Wide-area system monitoring and control, *IEEE Power & Energy Magazine*, 68–76, September/October 2004.

Katz, O., Talmon, R., Lo, Y. L., Wu, H. T., Alternating diffusion maps for multimodal data fusion, *Information Fusion*, 45, 346–360, 2019.

Khaleghi, B., Khamis, A., Karray, F. O., Razavi, S. N., Multisensor data fusion: A review-of-the state of the art, *Information Fusion*, 14, 28–44, 2013.

Kincic, S., Wangen, B., Mittelstadt, W. A., Fenimore, M., Cassiadoro, M., Van Zandt, V., Pérez, L., Impact of massive synchrophasor deployment on reliability coordinating and reporting, *2012, IEEE Power and Energy Society, General Meeting*, San Diego, CA, July 2012.

Lee, L. A., Centeno, V., *Comparison of μPMU and PMU, 2018 Clemson University Power Systems Conference (PSC)*, Charleston, South Carolina, September 2018.

Messina, A. R., *Wide-Area Monitoring of Interconnected Power Systems*, IET Power and Energy Series 77, London, UK, 2015.

North American Electric Reliability Corporation (NERC), Real-time application of synchrophasors for improving reliability, Princeton, NJ, October 2010.

Pan, S., Morris, T., Adhikari, U., Classification of disturbances and cyber-attacks in power systems using heterogenous synchronized data, *IEEE Transactions on Industrial Informatics*, 11(3), 650–662, 2015.

Phadke, A. G., Thorp, J. S., *Synchronized Phasor Measurements and Their Applications*, Springer, New York, 2008.

Phadke, A. G., Thorp, J. S., Communication needs for wide-area measurement applications, *2010 5th International Conference on Critical Infrastructure (CRIS)*, Beijing, China, September 2010.

Pinte, B., Quinlan, M., Reinhard, K., Low voltage micro-phasor measurement unit (μPMU), *2015 IEEE Power and Energy Conference at Illinois (PECI)*, Champaign, IL, February 2015.

Roberts, C. M., Shand, C. M., Brady, K. W., Stewart, E. M., McMorran, A. W., Taylor, G. A., Improving distribution network model accuracy using impedance estimation from micro-synchrophasor data, *2016 IEEE Power Engineering Society General Meeting*, Boston, MA, July 2016.

Sacha, D., Zhang, L., Sedlmair, M., Lee, J. A., Peltonen, J., Weiskopf, D., North, S., Keim, D. A., Visual interaction with dimensionality reduction: A structured literature analysis, *IEEE Transactions on Visualization and Computer Graphics*, 23(1), 241–250, 2017.

Stewart, E., Stadler, M., Roberts, C., Reilly, J., Arnold, D., Joo, J. Y., Data-driven approach for monitoring, protection and control of distribution system assets using micro-PMU technology, *CIRED – Open Access Proceeding Journal*, 2017(1), 1011–1014, 2017.

Thomopoulos, S. C. A., Sensor selectivity and intelligent data fusion, *1994 International Conference on Multisensor Fusion and Integration for Intelligent Systems*, Las Vegas, Nevada, 1994.

Von Meier, A., Stewart, E., McEachern, A. et al., Precision micro-synchrophasors for distribution systems: A summary of applications. *IEEE Transactions on Smart Grid*, 8(6), 2926–2936, 2017.

Yang, L., He, M., Zhang, J., Vittal, V., *Spatio-Temporal Data Analytics for Wind Energy Integration (Springer Briefs in Electrical and Computer Engineering)*, Springer, Cham, Switzerland, 2014.

Zhang, G., Hirsch, P., Lee, S., Wide area frequency visualization using smart client technology, *2007 IEEE Power Engineering Society General Meeting*, Tampa, FL, June 2007.

Zhang, P., Li, F., Bhatt, N., Next generation monitoring, analysis, and control for the future smart control center, *IEEE Transactions on Smart Grid*, 1(2), 186–192, 2010.

Zhang, J., Vittal, V., Sauer, P., Networked information gathering and fusion of PMU data: Future grid initiative white paper, PSERC Publication 12-07, May 2012.

Zuo, J., Carroll. R., Trachian, P., Dong, J., Affare, S., Rogers, B., Beard, L., Liu, Y., Development of TVA SuperPDC: Phasor applications, tools, and event Replay, *2008 IEEE Power and Engineering Society General Meeting*, Arizona State University, Pittsburg, CA, July 2008.

2

Data Mining and Data Fusion Architectures

2.1 Introduction

In recent years, a number of power utilities have designed and implemented advanced wide-area monitoring systems (WAMS) to enhance grid stability and reliability. Progress in wide-area monitoring has been enabled by advances in sensor, communication, and computing technologies. Advances in storage and related technologies, in turn, have resulted in massive, complex data sets that must be integrated to reduce uncertainty and enhance situational awareness.

In wide-area applications, each sensor (from its individual location) may capture both common and unique information about an observed phenomenon, yet it only provides a partial view of global behavior. Some of these dynamic processes manifest themselves at large scales (slow phenomena), while others (as in the case of electromagnetic transients) are dominated by fast fluctuations. An illustrative example is that of wide-area monitoring systems, where a properly designed *array* of sensors may provide a better description of slow global phenomena than would be gained from a single sensor alone. A key problem is how to reduce the number of measurements while retaining important information inherent to the data. These considerations clearly motivate the need for a joint analysis of the data.

Data fusion and data mining techniques can be used to address various issues in power system health monitoring and help support decision making. These include extraction of data and signal features and damage detection diagnostics, which may serve to predict future performance. Global analysis methods take advantage of cross information and may be used to transform data into knowledge using advanced data analysis tools (Sinha 2006). Several issues, however, make the mining and fusion of power system measured data difficult:

- Power system and data characteristics have evolved from centralized configurations to decentralized networks. Mining of distributed data is especially challenging as not all the data is readily available for analysis (due, for instance, to privacy concerns). Further, data characteristics in many fields have grown from static to dynamic and spatiotemporal, as seen in the case of wide-area monitoring systems (Chu et al. 2014).

- Modern PMUs provide multimodal or diverse data that must be integrated. This data may be highly correlated (as is often the case with nearby sensors or signals with similar characteristics) or the data may exhibit complementary features. Redundancy can make the extraction of useful information difficult and lead to poor performance of data fusion and data mining techniques.

- Power system time series coming from PMUs and other dynamic recorders are high-dimensional and noisy. With recent advances in measurement and storage capabilities, record lengths can be very large; this makes the application of some techniques difficult which may result in degraded performance.

- Missing and uncertain data, noise, and other artifacts may affect the performance of data fusion algorithms or lead to inaccurate predictions. In many applications, sensors are unevenly distributed resulting in oversampling or poor network coverage.

- In parallel with the growth in system complexity, new generation resources are being installed resulting in highly distributed data mining/data fusion issues.

There are a number of different concepts that contribute to the definitions of *data mining* and *data fusion*. Broadly speaking, *data mining* can be defined as the process of extracting unknown patterns and relationships in large data sets to predict future behavior (Khan et al. 2008; Hall and Llinas 1997). Thus, for instance, intelligent methods for automated pattern analysis can be used to monitor system behavior, detect cyberphysical attacks, predict impending events, and detect emerging trends. Moreover, data mining techniques can be used to mine and classify disturbance events recorded using WAMS (Dahal and Brahma 2012). Subtasks include spatiotemporal forecasting, spatiotemporal association rule mining, spatiotemporal sequential pattern mining, and spatiotemporal clustering and association, among others. *Data fusion*, on the other hand, refers to the integration of various types of data to obtain more information from the combined data instead of considering each data set separately. It can be used to generate new (fused) data, improve the data output quality, extract knowledge from the data, or make decisions regarding the state of the system.

Data fusion techniques lie at the heart of modern smart meters, smart grid integration, and other applications (Lau et al. 2019; Almasri et al. 2015). There is a common consensus that access to and fusion of heterogeneous, multimodal data may provide a more complete representation of the underlying phenomena being studied. Advantages of multisensor data fusion include:

- Reduced uncertainty of the information,
- Improved noise rejection,
- Robustness and reliability,
- Reduced ambiguity,
- Enhanced spatial and temporal resolution of the data,
- Improved dimensionality, and
- Increased effective data rates or data volume.

As with other emerging technologies, multisensor data fusion has some disadvantages and limitations. First, there are no perfect, universal algorithms—each application requires a different approach. Moreover, the fused information may be difficult to interpret or understand.

Figure 2.1 shows a generic data mining/data fusion architecture. In a typical setup, multichannel PMUs or dynamic sensors provide streams of data that need to be fused and analyzed. The information fusion can typically occur at three different levels: (1) Raw level, (2) Feature level, and (3) Global level.

In parallel with the growth in data complexity there are other factors that make the need for data fusion more significant.

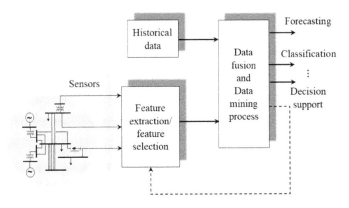

FIGURE 2.1
Data mining/data-fusion architecture incorporating information from sensors and other sources. (Based on Raheja et al., 2006.)

2.2 Trends in Data Fusion and Data Monitoring

Most traditional analysis methods address data fusion and data mining techniques as two separate problems. These techniques, however, are interrelated and interdependent in various ways.

It is only recently, that power system data fusion architectures and algorithms have begun to be developed and applied that monitor power system behavior. Efforts to incorporate data fusion and data mining techniques to power system monitoring are described in several recent works (NASPI 2019; Zhang et al. 2012; Messina 2015; Messina et al. 2014; Fucso et al. 2017; Usman and Faruque 2019). These efforts, however, are often application-specific and do not provide an integrated approach to power system monitoring.

The importance of fusing multimodal data has been pointed out in several research publications. Significant impact areas include:

- The determination of states and devices to be monitored—another related question is which combination of joint features is optimal in each project.
- The analysis of warning signals of impending failures.

At the heart of data mining techniques are algorithms that look for patterns in the data that can be described or classified. This description or classification then becomes knowledge which can then be turned into rules for analysis or prediction.

2.3 Data Mining and Data Fusion for Enhanced Monitoring

Current data fusion architectures and algorithms are largely based on abstractions of applications from various fields (Hall and Llinas 1997; Llinas and Hall 1998; Raheja et al. 2006). Such fields of application include condition-based maintenance (Raheja et al. 2006; Lahat et al. 2015), smart-city applications (Lau et al. 2019), wide-area monitoring (Messina 2015), structural health monitoring (Wu et al. 2018), and biometric recognition (Haghighat et al. 2016).

Figure 2.1 is an illustration of a data fusion/data mining architecture showing monitoring points. The model consists of four major modules or levels:

1. Data acquisition and cleansing (pre-processing),
2. Feature extraction and feature selection,
3. Data fusion and data mining,
4. Decision support.

At the same time new data fusion architectures are being developed to facilitate the reduction of wide bandwidth sensor data.

2.3.1 Data Fusion

Data fusion can be defined as the process of combining data from different sources (sensors or PMUs) to provide a more robust and complete description of an environment or process of interest (Hall and Llinas 1997). At the core of these processes are advanced analytical techniques used to process massive amount of data in near real-time (Lahat et al. 2015; The National Academies of Sciences 2008; Lopez and Sarigul-Klijn 2010).

Data fusion is a key enabler for enhanced situational awareness and power system health monitoring. Major advantages resulting from the use of data fusion techniques include (Castanedo 2013):

- Enhancement of data quality such as improved signal-to-noise ratios,
- Increased robustness and reliability in the event of sensor failure,
- Extended parameter coverage, rendering a more complete picture of the system,
- Reduction in data redundancy, and
- Improved consistency and reduced uncertainty.

As a by-product, the output of a data fusion algorithm is a more consistent, accurate, and reliable data set. Reliable data plays a central role in the analysis, monitoring, and forecasting of system behavior. Also, consistency is of interest since different measurements or data values may result in conflicting information.

2.3.2 Time-Series Data Mining

Data mining is the process of turning raw data into useful information (NASPI 2019) and can be categorized into three main classes: (a) dimensionality reduction algorithms, (b) similarity measures, and (c) data mining techniques.

Standard data mining includes tasks such as:

- Dimensionality reduction algorithms,
- Association rule mining and sequential pattern mining,
- Clustering,
- Classification,
- Anomaly detection, and
- Regression and prediction and forecasting (Taniar 2008; Cheng et al. 2013; Dahal and Brahma 2012).

Data mining and data fusion, however, are evolving concepts that change as new algorithms and increased computer power are developed. Consequently, there is a clear need for well-founded robust and efficient methods for the analysis of multidimensional data.

The increasing complexity of the power grid due to higher penetration of distributed resources and growing availability of interconnected, distributed heterogeneous metering devices requires new tools and technologies for providing a unified, consistent view of the system (Fusco et al. 2017; Dahal and Brahma 2012).

2.4 Data Fusion Architectures for Power System Health Monitoring

In recent years, several multisensor data mining/data fusion models and architectures for monitoring observational data have been developed in different contexts and application fields (Katz et al., 2019). Castanedo (2013) describes generic structures that are often adapted to various applications in different domains. However, many of these strategies have been developed for static data and are of limited interest to power system applications.

Architectures can be classified based on a variety of attributes or features. A typical data fusion architecture is often organized into hierarchical layers and includes several levels of organization. According to the nature of the input data, fusion can be performed at three processing levels: (1) Raw data level, (2) Feature level, and (c) Decision level. One research team led by Dalla Mura (2015) explains these levels as:

1. *Raw data level (Low-level fusion)*: Signal-level (raw data) fusion. It is generally recognized that fusing multisensor data at the raw-data level may not always yield the best inference. Further, sensors may have a limited computational capability.

2. *Middle level data (feature) fusion*: The middle-level data fusion center is expected to perform data fusion over a geographical area or distribution network.

3. *High-level data fusion*: At this central level, data from different fusion centers is fused to provide a global estimate of system behavior.

Feature level fusion is often recognized as more effective for data fusion since it is expected to contain richer information about the input data or the physical process of interest; for example, oscillation monitoring (Messina 2015; Haghighat 2016). Also, combined data (including, for instance, voltage and frequency signals) may be more informative than single, unimodal data (Dutta and Overbye 2014; Ghamisi et al. 2019; Li et al. 2015).

Several conceptualizations of data fusion exist in the literature that might be of interest for power system applications. Figure 2.1 gives a conceptual illustration of simple data fusion architecture. These illustrations, however, do not consider data relationships and types and are therefore not suitable for application to power system data (Zhang et al. 2012).

Structures used in multisensor data fusion may be of the centralized, decentralized, and distributed types. Distributed and hierarchical structures are of special interest for power system monitoring for various reasons: (a) There is a requirement for information at regional as well as global scales, (b) Fusion architectures can be integrated to existing local power data concentrators, and (c) Sensors in the distributed fusion structure can be independent from each other and potentially heterogeneous. Smart sensors capable of acquiring data, extracting features, and performing sensor-level data fusion and pattern recognition are expected to be a part of the future smart grid.

At the next layer in the network, sensors collect information from several component health monitors. The highest level of the network coordinates and fuses the information from different system health monitors and provides a connection for human user interfaces to the system.

Inputs to the data fusion architecture are historical data and measured data collected from various dynamic recorders. Raw data is collected and delivered to the WAMS for processing using a number of dynamic recorders or sensors.

2.4.1 Centralized Data Fusion Architectures

Traditional data fusion architectures are centralized in nature (measurements from multiple sensors are processed globally in a single central node or filter for fusion) and the results are distributed to various users.

Figure 2.2 is an illustration of a centralized architecture. The resulting architectures are conceptually simple and more accurate but are inefficient when dealing with a large number of measurements and limited bandwidth. Furthermore, the performance of these architectures may be affected by communication delays and result in highly complex communication structures.

Central to these approaches is the need for high processing capabilities and wide communication bandwidths as transmitting large volumes of data may take up a substantial part of the available data fusion bandwidth. These architectures share common communication infrastructure issues and control strategies with centralized WAMS (Shahraeini et al. 2011). In turn, decentralized data fusion involves local sensing and filtering, and assimilation over a network-centric system. The local fusion nodes, however, are not systematically coordinated, which may result in deviations from optimal performance.

The growing increase in distributed generating resources along with trends toward interconnected systems and the development of more sophisticated sensors makes the study of other architectures necessary.

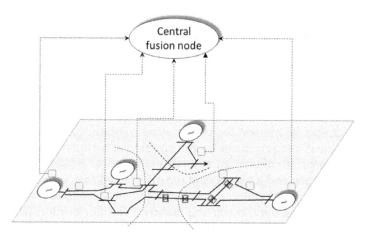

FIGURE 2.2
Centralized data fusion architecture. (Based on Raheja et al., 2006.)

A promising alternative that has recently emerged in a number of applications is to use distributed or hierarchical data fusion architectures based on reduced information structures (Cabrera et al. 2017). The following subsections briefly review distributed and hierarchical architectures in the context of their application to power system monitoring.

2.4.2 Distributed Data Fusion Architectures

A distributed data fusion system consists of a network of sensor nodes (filters) together with advanced data fusion algorithms. In distributed fusion architectures (Cabrera et al. 2017; Grime et al. 1994), data from multiple sensors are processed independently at individual fusion nodes to compute local estimates *before* they are sent to the central node for global fusion to produce the final state estimates. The information shared by the local fusion centers can be used to find consensus between subsystems.

Conceptually, the distributed data fusion structure consists of an ascending processing hierarchy in which local PDCs or fusion centers can be utilized to store and process information at low (local) levels (Messina 2015). Since the sensors independently process their associated measurements this structure may reduce communication delays and data communication errors between (local) fusion systems.

These estimates are then transmitted to other fusion nodes and to the central node to form a global state estimate and then are submitted for various applications as depicted in Figure 2.2. Compared to a centralized architecture, a distributed network of sensors is superior in many ways. In particular, distributed networks have the potential to solve problems in a cooperative manner, cover larger areas, and handle considerable increases in spatial resolution. In addition, local processing of the data means a lower processing load on individual nodes, lower communication costs, flexibility to respond to changes, and robustness to overcome point failures.

These architectures do not require any central fusion or central communication facility and offer some advantages when compared with the centralized counterparts. These include lighter processing load at each processing or fusion node, lower communication load, faster access to fusion results, and the avoidance of a large centralized data base (Liggins et al. 1997).

Figure 2.3 gives a conceptual representation of this type of architecture.

When the nodes communicate, each node fuses or integrates the information received from other nodes with the local information using some communication strategy to update situation or state assessment. Particular cases involving use of this model to monitor system behavior are described in Cabrera et al. (2017).

2.4.3 Hierarchical Data Fusion Architectures

In recent years, hierarchical and hybrid data fusion architectures, which combine the properties of distributed and decentralized architectures, have proved to be a useful alternative to centralized architectures. In this architecture, the data fusion step is performed at a variety of levels in the hierarchy. A conceptual illustration of a hierarchical structure is shown in Figure 2.4. Here, each node or data concentrator processes the data from its own set

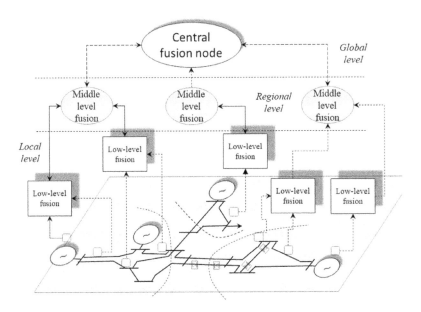

FIGURE 2.3
Schematic of a distributed data fusion architecture. (Based on Liggins II, M.E. et al., 1997.)

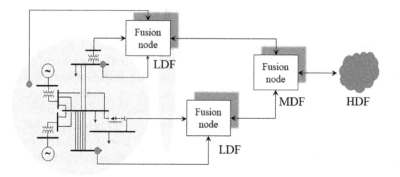

FIGURE 2.4
Multi-level hierarchical data fusion architecture. (Based on Liggins II, M.E. et al., 1997.)

of sensors and communicates with other nodes to improve on the overall estimates. As with the distributed fusion architecture, each PDC has its own local processor that can generally extract useful information from the new sensor data prior to communication. Various types of sensor measurements are analyzed separately. However, fusing multisensor data at the raw-data level may not always yield the best inferences.

2.4.4 Next-Generation Data Fusion Architectures

Next-generation data fusion systems must effectively integrate data from heterogeneous sensors (e.g., distribution sensors, frequency dynamic recorders and PMUs) to enhance situational awareness. Advanced data fusion systems should also be well suited to detect cyberspace intrusion and operate in a more uncertain operating scenario.

2.5 Open Issues in Data Fusion

This section introduces the problems of data fusion in the context of modern monitoring architectures used in various application domains (Dalla Mura et al. 2015; Lahat et al. 2015).

> *Redundancy*: Data fusion can eliminate or reduce redundant data and thus result in more efficient wide-area monitoring of the system. In practice, however, some degree of redundancy may be necessary to facilitate the detection and isolation of faulty sensors and to ensure accessibility of synchrophasor data in the event of failure of sensors or communications links (Zhang et al. 2012).

Further, redundant measurements may lead to inefficient performance of conventional diffusion methods.

Dimensionality reduction: Dimensionality reduction has become a popular approach for decreasing computational burden as well as for finding features that are not evident in the original input space. In the analysis and multisensor information, it is often the case that the observed variables are unknown functions of only a few independent sensors. Further, the potentially large number of sensor and fusion nodes poses significant challenges to data transmission and association.

Data concatenation: Simply concatenating two or more feature vectors into a single feature vector raises several numerical and conceptual problems. Both serial and parallel data concatenation methods have been used in this practice. More complex representations include multiblock fusion techniques. In practical applications, the use of heterogeneous data may require some degree of transformation or scaling to make analysis consistent.

Communication bandwidth: Constraints on communication bandwidth may severely affect the performance of data mining and data fusion structures. Recently, distributed data mining/data fusion approaches are gaining attention in system monitoring for improving data analysis and enabling the analysis of massive volumes of data. They are used to improve the reliability of delivered information. While this approach addresses data accuracy, it does not address the inefficiencies caused by very large nodes and high data redundancy.

Information (feature-level) fusion: Modern data fusion centers incorporate feature-level fusion for monitoring and recognition. A related issue is that of separating the classes of information by combining and processing information from two or more feature vectors or sources using some kind of correlation analysis (Haghighat et al. 2016). Accounting for associations across space and time, however, remains an open challenge for research.

Other issues include:

- Disparity of time scales,
- Data correlation, and
- Missing data and uncertainty.

References

Almasri, M., Elleithy, K., Data fusion in WSNs: Architecture, taxonomy, evaluation of techniques, and challenges, *International Journal of Scientific & Engineering Research*, 6(4), 1620–1636, 2015.

Cabrera, I. R., Barocio, E., Betancourt, R. J., Messina, A. R., A semi-distributed energy-based framework for the analysis and visualization of power system disturbances, *Electric Power Systems Research*, 143, 339–346, 2017.

Castanedo, F., A review of data fusion techniques, *The Scientific World Journal*, 6, Article 704504, 2013.

Cheng, T., Haworth, J., Anbaroglu, B., Tanaksaranond, G., Wang Jiaquiu, Spatio-temporal data mining, in Fischer, M. M., Nijkamp, P. (Eds.), *Handbook of Regional Science*, Springer Verlag, Berlin, Germany, 2013.

Chu, W. W., *Data Mining and Knowledge Discovery for Big Data—Methodologies, Challenge, and Opportunities*, Springer-Verlag, Berlin, Germany, 2014.

Dahal, O. P., Brahma, S. M., Preliminary work to classify the disturbance events recorded by Phasor measurements units, *IEEE Power and Engineering Society General Meeting*, San Diego, CA, July 2012.

Dalla Mura, M. D., Prasad, S., Pacifici, F., Gamba, P., Benediktsson, J. A., Challenges and opportunities of multimodality and data fusion in remote sensing, *Proceedings of the IEEE*, 103(9), 11(4), 1585–1601, 2015.

Dutta, S., Overbye, T., Feature extraction and visualization of power system transient stability results, *IEEE Transactions on Power Systems*, 29(2), 966–973, 2014.

Fusco, F., Tirupathi, S., Gormally, R., Power systems data fusion based on belief propagation, *2017 IEEE PES Innovative Smart Grid Technologies Conference Europe (ISGT-Europe)*, Torino, Italy, September 2017.

Ghamisi, P., Rasti, B., Yoyoka, N., Wang, Q., Hofle, B., Bruzzone, L., Bovolo, F. et al., Multisource and multitemporal data fusion in remote sensing, *IEEE Geoscience and Remote Science Magazine, IEEE Geoscience and Remote Sensing Magazine*, 6–39, March 2019.

Grime, S., Durrant-Whyte, H. F., Data fusion in decentralized sensor networks, *Control Engineering Practice*, 2(5), 849–863, 1994.

Haghighat, M., Abdel-Mottaleb, M., Alhalabi, A., Discriminant correlation analysis: Real-time feature level fusion for multimodal biometric recognition, *IEEE Transactions on Information Forensic and Security*, 11(9), 1984–1996, 2016.

Hall, D. L., Llinas, J., An introduction to multisensory data fusion, *Proceedings of the IEEE*, 85(1), 6–23, 1997.

Katz, O., Talmon, R., Lo, Y. L., Wu, H. T., Alternating diffusion maps for multimodal data fusion, *Information Fusion*, 45, 346–360, 2019.

Khan, S., Ganguly, A. R., Gupta, A., Data mining and data fusion for enhanced decision support, Chapter 27, in Burstein, F., Holsapple, C. W. (Eds.), *Handbook on Decision Support Systems*, pp. 581–608, Springer-Verlag, Berlin, Germany, 2008.

Lahat, D., Adali, T., Jutten, C., Multimodal data fusion: An overview of methods, challenges and prospects, *Proceedings of the IEEE*, 103(9), 1449–1477, 2015.

Lau, B. P. L., Marakkalage, S. H., Zhou, Y., Hassan, N. U., Yuen, C., Zhang, M., Tan, U., A survey of data fusion in smart city applications, *Information Fusion*, 52, 357–374, 2019.

Li, Y., Li, G., Wang, Z., Han, Z., Bai, X., A multifeatured fusion approach for power system transient stability assessment using PMU data, *Mathematical Problems in Engineering*, 2015, 786396, 2015.

Liggins, II, M. E., Chong, C. Y., Kadar, I., Alford, M. G., Vannicola, V., Thomopoulos, S., Distributed fusion architectures and algorithms for target tracking, *Proceedings of the IEEE*, 85(1), 95–107, 1997.

Llinas, J. and Hall, D. L., An introduction to multi-sensor data fusion, in *Proceedings of the International Symposium on Circuits and Systems*, Monterey, CA, May–June 1998.

Lopez, I., Sarigul-Klijn, N., A review of uncertainty in flight vehicle structural damage monitoring, diagnosis and control: Challenges and opportunities. *Progress in Aerospace Sciences*, 46, 247–273, 2010.

Messina, A. R., *Wide-Area Monitoring of Interconnected Power Systems, Power and Energy Series 77*, IET, Stevenage, UK, 2015.

Messina, A. R., Reyes, N., Moreno, I., Perez G., M.A., A statistical data-fusion-based framework for wide-area oscillation monitoring, *Electric Power Components and Systems*, 42(3–4), 396–407, 2014.

NASPI white paper: Data mining techniques and tools for synchrophasor data, Prepared by NASPI Engineering Analysis Task Team (EATT), PNNL-28218, January 2019.

Raheja, D., Llinas, J., Nagi, R., Romanowski, C., Data-fusion/data mining-based architecture for condition-based maintenance, *International Journal of Production Research*, 44(14), 2869–2887, 2006.

Shahraeini, M., Javidi, M. H., Ghazizadeh, M. S., Comparison between communication infrastructures of centralized and decentralized wide area measurement systems, *IEEE Transactions on Smart Grid*, 2(1), 206–211, 2011.

Sinha, A. K. (Ed.), *Geoinformatics: Data to Knowledge*, The Geological Society of America, Boulder, CO, 397, January 2006.

Taniar, D., *Data Mining and Knowledge Discovery Technologies*, IGI Publishing, Covent Garden, London, UK, 2008.

The National Academies of Sciences, Engineering, and Medicine, In-time aviation safety management: Challenges and research for an evolving aviation system, A Consensus Study Report, The National Academies Press, 2018, https://doi.org/10.17226/24962.

Usman, M. U., Faruque, M. O., Application of synchrophasor technologies in power systems, *Journal of Modern Power Systems and Clean Energy*, 7(2), 211–226, 2019.

Wu, R. T., Jahanshani, M. R., Data fusion approaches for structural health monitoring and system identification: Past, present, and future, *Structural Health Monitoring*, 1–35, 2018.

Zhang, J., Vittal, V., Sauer, P. Networked information gathering and fusion of PMU data: A broad Analysis Prepared for the Project the future grid to enable sustainable energy systems, Power Systems Engineering Research Center, PSERC Publication 12-07, May 2012.

Section II

Advanced Projection-Based Data Mining and Data Fusion Techniques

3

Data Parameterization, Clustering, and Denoising

3.1 Introduction: Background and Driving Forces

Central to the development of advanced monitoring systems that improve the current predicting capabilities of wide-area monitoring systems is the investigation of data parameterization, denoising, and clustering in both space and time.

Fusing multiple sets of synchronized measurements for power system monitoring is a challenging and complex problem, as it often involves signals with differing units and disparate scales. Spatiotemporal measurements are sparsely distributed over a large geographical area and tend to exhibit trends and other artifacts that make the analysis of wide-area phenomena difficult. With new and more sophisticated dynamic recorders being deployed in the system, techniques to standardize inhomogeneous data sets are needed.

In analyzing large spatiotemporal processes several issues are of interest, including, filtering, detecting abrupt changes in the observed behavior, automated spatiotemporal clustering of trajectory data, and pattern recognition. Central to these approaches, is the ability to understand the nature of dynamic interactions and associations between dynamic trajectories or sensors.

Trajectory data mining has recently emerged as a key area of research for system monitoring (Messina 2015; Dutta and Overbye 2014; Rovnyak and Mei 2011). Data mining and data fusion techniques should be capable of efficiently modeling and analyzing massive data sets and extracting hidden patterns. Clustering trajectory data is a recent problem in the analysis of spatiotemporal data sets, with many contributions found in the data mining and physical sciences literature (Bezdek 1981; Bezdek et al. 1984; David and Averbuch 2012; Meila and Shi 2001; Scott et al. 2017).

There are currently several data characterization and clustering methods being used for power system monitoring analysis. Many existing approaches to data parameterization and clustering, however, provide unsatisfactory results and limited utility when applied to power system data. This chapter examines the issues of data characterization, data clustering, and denoising in power system applications.

First, a framework for the analysis and characterization of spatiotemporal measured data is introduced within the realm of trajectory mining techniques. The notion of temporal distances is introduced, and techniques to measure distances are discussed. Drawing upon the notion of kernel-based dimensionality reduction techniques, methods to characterize the observed behavior in terms of spatial and temporal modes are introduced and tested on observational data. By incorporating concepts of information geometry, these methods extend naturally to the nonlinear case.

Promising methods for extracting the dominant temporal and spatial components of variability in measured data are also investigated and numerical issues are addressed.

3.2 Spatiotemporal Data Set Trajectories and Spatial Maps

3.2.1 Dynamic Trajectories

Measured dynamic data have a useful interpretation in terms of spatiotemporal trajectories that evolve with time. Following Messina and Vittal (2007), assume that $x_k(t_j)$ denotes a sequence of observations of a measured transient process at locations x_k, $k = 1, \ldots, m$, and time t_j, $j = 1, \ldots, N$, where the m locations represent sensors.

More formally, the time evolution of the transient process can then be described by the mxN-dimension observation (snapshot) matrix X,

$$X = \begin{bmatrix} x_1 & x_2 & \cdots & x_N \end{bmatrix} = \begin{bmatrix} x_1(t_1) & x_1(t_2) & \cdots & x_1(t_N) \\ x_2(t_1) & x_2(t_2) & \cdots & x_2(t_N) \\ \vdots & \vdots & \ddots & \vdots \\ x_m(t_1) & x_2(t_2) & \cdots & x_m(t_N) \end{bmatrix} \in \mathfrak{R}^{mxN}, \quad (3.1)$$

where $x_j = \begin{bmatrix} x_1(t_j) & x_2(t_j) \ldots & x_m(t_j) \end{bmatrix}^T, j = 1, \ldots, N$.

Figure 3.1 provides a spatiotemporal representation of data collected using a network of sensors. Both purely spatial and spatiotemporal representations are possible. In the first case, topological (spatial) information is provided from which directed graphs and spatial map patterns can be obtained (Borcard et al. 2002; Griffith 2000). In the second case, spatiotemporal information from a sensor network is used to extract spatiotemporal patterns (Kezunovic and Abur 2005; Messina and Vittal 2007; Barocio et al. 2015).

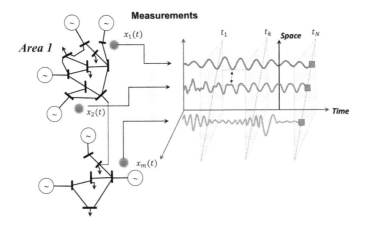

FIGURE 3.1
Sensor network showing time sequences of dynamic trajectories.

Constant monitoring of dynamic trajectories is of interest to detect deviations from normal expected operation, cluster dynamic trajectories, assess power system health, and develop early warning systems. At the core of spatiotemporal data mining approaches is the notion of distance between dynamic trajectories.

The most common analysis techniques for computing the distance of a dynamic trajectory or set of trajectories from a reference is the root mean square deviation. At each time instance, the instantaneous (pointwise) distance between two trajectories or time sequences $x_i(t)$ and $x_j(t_j)$, $t_j = 1, 2, \ldots, N$ is given by

$$d_{ij}(x_i, x_j, t_k) = \underbrace{\left\| x_i(t_k) - x_j(t_k) \right\|}_{\substack{\text{Distance} \\ \text{function}}}, \tag{3.2}$$

where $x_i(t_k)$ and $x_j(t_k)$ represent the instantaneous values of the time sequences at time instance t_j (a time slice in the spatiotemporal representation in Figure 3.1). The Euclidean distance, however, becomes negligible beyond the local neighborhood of each data point and is sensitive to noise. Further, Euclidian distances may fail to characterize nonlinear manifolds. This has motivated the development of more reliable distance functions.

More general measures of distances can be obtained using the notion of *windowing*. For a time interval $\{t_1, t_2, \ldots, t_N\}$ the distance between trajectories $x_i(t), x_j(t)$ over time becomes

$$d_{ij}(x_i, x_j, t_k) = \sqrt{\frac{1}{n} \sum_{k=1}^{n} \left\| x_i(t_k) - x_j(t_k) \right\|}, \tag{3.3}$$

where $x_i(t_k)$ is the position or value of $x_i(t)$ at time t_k.

Distances can be measured with respect to a reference frame (either fixed or moving) or a mean value. Of particular interest within this realm are root mean squares fluctuations, since they allow the analysis of deviations from reference values.

3.2.2 Proximity Measures

To introduce and motivate the need for data trajectory mining consider the case where all trajectories are finite in length, and there are no missing observations.

Suppose that time histories of selected variables $x_k(t_j)$, $k=1,\ldots,m$, $j=1,\ldots,N$, where N is the number of observations at m system locations or sensors, are simultaneously recorded using homogeneous sensors. At each time step, a distance or proximity measure, $d_{ij}(t_j)$, between two trajectories, $x_i, x_j, i=1,\ldots,m, j=1,\ldots,m$ can be defined as suggested in Equation (3.2) and illustrated in Figure 3.2.

Several measures of affinities or trajectories have been introduced in the literature ranging from Euclidean to cosine-type similarities. Table 3.1 summarizes some common distance metrics used in the power systems literature.

Viewing the data sets as points in the data space, a distance or similarity matrix describing the proximity or similarity between objects (trajectories) can be defined as

$$S = \begin{bmatrix} s_{ij} \end{bmatrix} = \begin{bmatrix} s_{11} & \cdots & s_{1m} \\ \vdots & \ddots & \vdots \\ s_{m1} & \cdots & s_{mm} \end{bmatrix}, \tag{3.4}$$

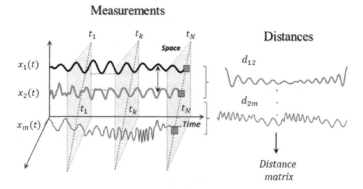

FIGURE 3.2
Dynamic trajectories for spectral dimensionality reduction.

TABLE 3.1

Typical Similarity or Distance Functions

Model	Feature
Euclidean (distance-based)	$\left\| x_i(t) - x_j(t) \right\|^2$
Cosine affinity	$\dfrac{x_i^T(t) x_j(t)}{\sqrt{\left(x_i^T(t) x_i(t) \right) \left(x_j^T(t) x_j(t) \right)}}$
Squared Mahalanobis distance	$\left(x_i(t) - x_j(t) \right) \Sigma \left(x_i(t) - x_j(t) \right)^T$

where the coefficients s_{ij} provide a measure of similarity between trajectories (Borcard et al. 2002). Distance matrices are full, symmetric, and diagonally dominant.

Typically, the similarity measure becomes negligible beyond the local neighborhood of each data vector. As a result, large Euclidean distances do not provide useful similarity information. This is related to the notion that on a general nonlinear manifold, large Euclidean distances cannot be expected to reliably approximate geodesic distances.

To pursue this idea further let the distance (similarity measure) between two time trajectories $x_i(t)$, $x_j(t)$, $i, j = 1, \ldots, m$ up to time t be defined as

$$d_{ij}(t) = \left\| x_i(t) - x_j(t) \right\|^2 = \sum_{k=1}^{m} \left(x_{ki}(t) - x_{kj}(t) \right). \tag{3.5}$$

Generalizations to this model to define local instantaneous distances between trajectories are described in subsequent chapters.

Intuitively, coherent trajectories or data sets are tight bundles of trajectories, such that $s_{ij}(t) = \left\| x_i(t) - x_j(t) \right\|$ is small for all i, j; the objective is to integrate these trajectories as part of a coherent set.

A few comments are in order:

- The measurement points (sensor locations) and their relations form a network (or graph) described by weights given by the Euclidean distances that indicate the existence of a relation between point i and point j.

- Alternatively, a distance matrix, S, can be defined as a matrix containing zeros when the trajectories are very close ($s_{ij} = 0$), and values s_{ij} given by the pairwise Euclidean distance coefficients when $x_i \neq x_j$.

- Information about time is lost, which makes physical interpretation of dynamic processes difficult and motivates the need for the development of more effective distance measures.

3.2.3 Spectral Graph Theory

Similarity or distance matrices have a natural interpretation in terms of weighted, undirected graphs[1]. From Figure 3.2, a weighted graph $G(V,w)$, with m vertexes $V = \{v_1 \quad v_2 \ldots \quad v_m\}$ and weights $w_{ij} = w(v_i, v_j)$, can be constructed by computing a pairwise similarity distance between trajectories. Each sensor location corresponds to a vertex, while the edges (weights) represent the interaction strength or affinity between trajectories (Newman 2018). See Figure 3.3.

Using the distance between motions or trajectories, the weight on the graph edges is given by

$$w_{ij} = \left\| x_i - x_j \right\|,$$

where

$$w_{ii} = \sum_{j=1}^{N} w_{ij}.$$

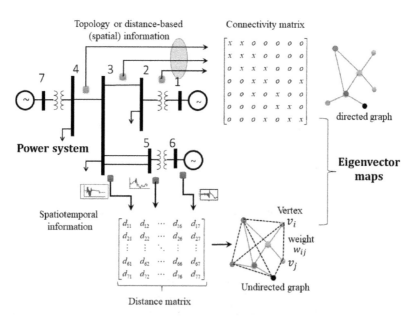

FIGURE 3.3
Illustration of distance matrices and associated undirected graphs.

---

[1] The graph or network is *undirected* if $w_{ij} = w_{ji}$ for all ij, otherwise it is *directed* (Newman 2010).

Each measurement or data point can be associated with a different neighborhood. It follows that the sum of the squared distances of all points, x_j, within cluster C_j to its centroid c_j, is given by

$$SST = \sum_{j=1}^{m} \left(x_j - c_j\right)^2. \tag{3.6}$$

The SST index lies between 0 and 1 and gives the percentage of explained variance by the data and is similar to the $R2$ value in regression analysis.
Several remarks are in order:

- The calculation of w_{ij} requires consideration of all possible pathways on the system (by definition w_{ii} is zero).
- Without a proper mapping, different data sets cannot be compared directly.
- The high-dimensional representation may contain a great deal of redundancy.
- The graph definition does not account for the time ordering of the signals.

Vertex Eccentricity: The maximum value entry in the ith row of the distance matrix is called *vertex eccentricity* and is given by

$$\eta_i = \max_j \left(d_{ij}\right).$$

In Section 3.3, a brief introduction to distance-based clustering techniques is presented.

3.3 Power System Data Normalization and Scaling

Data recorded by different sensors may have different scales, units, and complementary features, which makes direct comparison meaningless. Thus, for instance, sensors placed in some regions may capture local oscillations while others may provide complementary information on global phenomena. Further, multisensor data may exhibit highly correlated features or redundancy associated with placement locations.

Data normalization may help to improve the physical content of measurements and allow joint analysis of different types of measurements (multiclass data) (Van den Berg et al. 2006; Bro and Smilde 2003).

3.3.1 Departure from Mean Value

In several applications, especially those associated with oscillatory processes, the measured data is written as the sum of a time-varying mean and a fluctuating component.

Consider a data set of the form (3.1), $X = [x_1 \quad x_2 \quad \cdots \quad x_N] \in \Re^{m \times N}$. The temporal behavior of each time sequence can be expressed in the form

$$x_k(t) = \sum_{j=1}^{p} x_j = m_k(t) + \sum_{j=1}^{p} c_k + e_k, \quad k = 1, \ldots, m, \tag{3.7}$$

where $m_k(t)$ is the time-varying mean, and the c_k are fluctuating components.

Following Messina (2015), let the time-varying means for each time series be stacked in the feature vector, x_{mean}, as

$$x_{\text{mean}} = \begin{bmatrix} m_1(t) & m_2(t) \ldots & m_m(t) \end{bmatrix}^T, \tag{3.8}$$

where the $m_k(t)$ represent the instantaneous means.

Deviations from the mean value can be written in compact form as

$$\hat{X} = X - 1_m x_{\text{mean}}^T = HX, \tag{3.9}$$

where 1_m is a vector of dimension N with all elements unity, T denotes the transpose operation, and H is a centering matrix of dimension $m \times m$, defined as

$$H = \left(I_m - \frac{1}{N} 1_m 1_m^T \right),$$

where I_m denotes an $m \times m$ identity matrix (Bro and Smilde 2003).

Using this approach, each element of the measurement matrix (3.1) can be expressed as

$$\hat{x}_{ij} = x_{ij} - m_i = x_{ij} - \frac{1}{N} \sum_{i=1}^{N} x_{ij},$$

where $\hat{x}_{ij}$ is the ijth element of the centered matrix. Alternatively centering can be performed using demeaning analysis techniques such as wavelet analysis (Baxter and Upton 2002), dynamic harmonic regression (Young et al. 1999; Zavala and Messina 2014), and the Hilbert-Huang technique (Huang 1998) to mention a few.

As an example, consider the application of detrending techniques to the data sets in Figure 3.4. Figure 3.5 illustrates the application of centering to simulated data using two different analytical approaches:

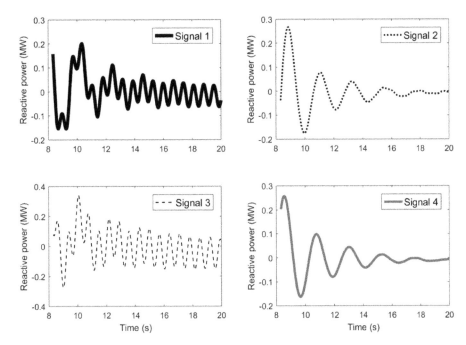

FIGURE 3.4
Simulated data for illustration of detrending techniques.

1. Centered data using (3.9), and
2. Centered data using the dynamic harmonic regression framework found in Zavala and Messina (2014).

Results are found to be in good agreement, although nonlinear techniques result in more symmetrical detrended signals which allows for better characterization of oscillatory phenomena. The use of other detrending and denoising techniques is discussed in Chapter 8.

3.3.2 Standardization

Data standardization is of particular interest when attempting to compare and fuse multimodal data in different units. One of the most popular methods is *z-score*. In this case, the data is centered around 0 and is scaled with respect to the standard deviation as

$$\hat{x}_i = \frac{x_i - \mu(x_i)}{\sigma(x_i)}, \qquad (3.10)$$

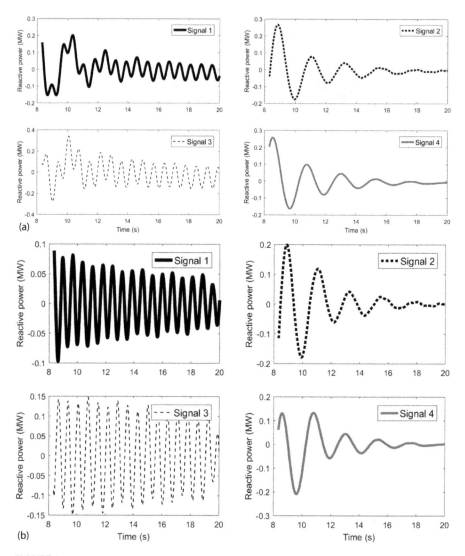

FIGURE 3.5
Centered and detrended data in Figure 3.4. (a) Centered data. (b) Wavelet detrended data.

where μ and σ are the mean and the standard deviation of the data set (Bro and Smilde 2003). Hybrid approaches involving detrending with scaling techniques are also possible, but they are likely to increase nonlinearity in the signals.

There are several caveats to performing data standardization and scaling. Centering and scaling may destroy sparsity in the data, reduce model interpretability, and, in some cases, lead to spurious results, especially in the case of missing data.

3.4 Nonlinear Dimensionality Reduction

3.4.1 Diffusion Maps

An efficient way to define a measure of dynamic proximity between different trajectories is using diffusion maps (DMs) (Lafon et al. 2006; Arvizu and Messina 2016). Given a measurement matrix X in (3.1), the diffusion distance can be found by introducing a random walk on the data set to ensure that the data matrix is positive definite.

A positive-definite matrix kernel K can now be obtained whose (i,j)th element is given by

$$[K] = [k_{ij}] = \exp\left(-\frac{\|x_i(t) - x_j(t)\|}{\varepsilon_i \varepsilon_j}\right), 1 \leq i, j \leq m. \tag{3.11}$$

The transition probability p_{ij} from i to j can now be obtained as $p_{ij} = k_{ij} / \sum_{k=1}^{m} k_{ik}$. In a variation of this approach shown in Lindenbaum (2015), an affinity matrix between the feature vectors can be defined as

$$\tilde{K} = [\tilde{k}_{ij}] = \frac{x_i^T(t) x_j(t)}{\sqrt{\left(x_i^T(t) x_i(t)\right)\left(x_j^T(t) x_j(t)\right)}}, 1 \leq i, j \leq m.$$

In practice, the pairwise distances in (3.15) are combined with a Gaussian kernel of bandwidth ε to retain small *rms* values. This transformation yields the *m*-by-*m* matrix **K** (a transition probability kernel), with elements k_{ij}. The bandwidth scale, ε, modulates the notion of distance in (3.11). As discussed in Taylor et al. (2011), if $\varepsilon_i \varepsilon_j \gg \max_{ij} \|x_i(t) - x_j(t)\|$, then for all edges $\{i, j\}, k_{ij} \approx 1$.

Table 3.2 summarizes some common similarity or distance functions used in the literature.

Define now a diagonal matrix $D = [d_{ij}]$ whose entries are the row sums of K, namely

$$D = [d_{ij}] = \begin{bmatrix} \sum_{j=1}^{m} K_{1j} & 0... & 0 \\ 0 & \sum_{j=1}^{m} K_{2j} & 0 \\ 0 & 0... & \sum_{j=1}^{m} K_{mj} \end{bmatrix},$$

where $d_{ii} = \sum_{j=1}^{m} k_{ij}$ is the degree of node x_i.

TABLE 3.2

Typical Similarity or Distance Functions

Algorithm	Distance Metric
Squared Mahalanobis distance	$(x - \bar{x})\Sigma^{-1}(x - \bar{x})$,
Euclidean distance	$\left(x_i(t) - x_j(t)\right)\left(x_i(t) - x_j(t)\right)^T$,
Weighted Euclidean metric	$\tilde{K} = [\tilde{k}_{ij}] = \dfrac{x_i^T(t)x_j(t)}{\sqrt{\left(x_i^T(t)x_i(t)\right)\left(x_j^T(t)x_j(t)\right)}}$,
Cosine distance metric	$\tilde{K} = [\tilde{k}_{ij}] = \dfrac{x_i^T(t)x_j(t)}{\sqrt{\left(x_i^T(t)x_i(t)/w\right)/\left(x_j^T(t)x_j(t)/w\right)}}$,

A Markov transition matrix, M, can now be defined as

$$M = \left[m_{ij}\right] = D^{-1}K = \begin{bmatrix} \dfrac{k_{11}}{\sum_{j=1}^{m} K_{1j}} & \dfrac{k_{12}}{\sum_{j=1}^{m} K_{1j}} & \cdots & \dfrac{k_{1m}}{\sum_{j=1}^{m} K_{1j}} \\[2ex] \dfrac{k_{21}}{\sum_{j=1}^{m} K_{2j}} & \dfrac{k_{22}}{\sum_{j=1}^{m} K_{2j}} & \cdots & \dfrac{k_{2m}}{\sum_{j=1}^{m} K_{2j}} \\[2ex] \vdots & \vdots & \ddots & \\[2ex] \dfrac{k_{m1}}{\sum_{j=1}^{m} K_{mj}} & \dfrac{k_{m1}}{\sum_{j=1}^{m} K_{mj}} & \cdots & \dfrac{k_{2m}}{\sum_{j=1}^{m} K_{mj}} \end{bmatrix}. \tag{3.12}$$

The following properties of the Markov matrix can easily be verified:

- Matrix M is nonnegative, unsymmetrical, and row-stochastic (rows sum to 1), and
- Matrix M is invariant to the observation modality and is resilient to measurement noise.

The spectrum of the Markov matrix provides information on the distribution of energy as a function of scale and the separation of the dynamic patterns. Spectral analysis of a locally scaled kernel provides a convenient way to extract the low-dimensional structure prevalent in the data. The eigenvalue problem for the operator M can be defined as

$$M\psi_j = \sigma_j \psi_j,$$

where ψ_j are right singular vectors associated with σ_j with corresponding left eigenvectors φ_j; matrix M has a complete set of singular values σ_j, of decreasing order of magnitude $\sigma_o > \sigma_1 > ... \sigma_d > 0$. Clearly, if the solution of the system $M\psi_o = \sigma_o\psi_o$ yields with $\sigma_o = 1$, $\psi_o = [1 \quad 1... \quad 1]^T$, then, collecting all eigenvalues yields

$$MY = \sigma_j Y$$
$$M\Phi = \sigma_j \Phi,$$

(3.13)

with $Y = [\psi_o \quad \psi_1 \quad \cdots \quad \psi_{m-1}]$ and $\Phi = [\phi_o \quad \phi_1 \quad \cdots \quad \phi_{m-1}]$.

Several features of this model are worth emphasizing:

- The eigenvectors associated with the largest singular values correspond to slow modes governing the long-time system evolution.
- Numerical experience with large spatiotemporal models shows that $\psi_o\psi_o^T > \psi_1\psi_1^T > ... > \psi_d\psi_d^T$, and $\psi_i\psi_j^T \approx 0, i, j = 1, ..., d$.
- In practice, a spectral gap can be observed at σ_d, such that

$$\underbrace{\sigma_1 \geq \sigma_2 \geq ... \geq \sigma_d}_{\text{Slow motion}} \gg \sigma_{d+1} \geq ... \sigma_m.$$

It follows that

$$M = D^{-1}K = D^{-1/2}\left(D^{-1/2}KD^{1/2}\right)D^{-1/2} = D^{-1/2}M_sD^{-1/2},$$

(3.14)

where

$$M_s = D^{-1/2}KD^{-1/2} = D^{-1/2}DMD^{-1/2} = D^{1/2}MD^{-1/2}$$

is a normalized affinity matrix (a normalized kernel).

Since, matrices M and M_s are related by a similarity transformation, they share the same eigenvalues. To prove this, assume that ψ denote the eigenvalues of M and Φ denote those of M_s.

It then follows that the eigenvectors of M_s satisfy

$$(M_s - \lambda I)\phi = 0 = \left(D^{1/2}MD^{-1/2} - \lambda I\right)\phi$$

or

$$(M - \lambda I)D^{1/2}\phi = 0 = \left(D^{1/2}MD^{-1/2} - \lambda I\right)\phi,$$

from which it follows that $\Phi = \phi$, and M and M_s share the same eigenvalues.

3.4.1.1 Numerical Issues

Noting that $M_s = D^{-1/2}KD^{-1/2}$, it can be inferred that M_s is a symmetric matrix, which allows the use of special techniques for calculating the associated singular values.

It follows from linear analysis that M_s is symmetric and therefore diagonalizable and positive definite with a decomposition

$$M_s = D^{-1/2}MD^{-1/2} = U\Sigma U^T = D^{-1/2}MD^{-1/2} = D^{-1/2}U\Sigma U^T D^{-1/2}$$

or

$$M_s = \Psi\Lambda\Phi = \sum_{j=1}^{m}\sigma_j\psi_j\phi_j,$$

where $\psi = \begin{bmatrix} \psi_o & \psi_1 & \cdots & \psi_{m-1} \end{bmatrix}, \Phi = \begin{bmatrix} \phi_o & \phi_1 & \cdots & \phi_{m-1} \end{bmatrix}$.

The d-dimensional diffusion map is defined at time t, as the map:

$$\Psi = \begin{bmatrix} \sigma_1\psi_1 & \sigma_2\psi_2 \cdots & \sigma_d\psi_d \end{bmatrix}, \tag{3.15}$$

where d is the number of relevant coordinates. By truncating the spatial patterns, dimensionality reduction can be performed.

3.4.1.2 Diffusion Distances

Associated with the spectral model, the diffusion distance can now be defined in terms of the forward probabilities M as

$$D_{ij} = \sum_{r=1}^{m}\frac{\left(M_{ir} - M_{jr}\right)}{\Psi(x_r)}; \quad \Psi(x_m) = \frac{\sum_{j=1}^{m}M_{jm}}{\sum_{k=1}^{m}\sum_{j=1}^{m}M_{jk}}$$

Physically, the diffusion distance is small if there are many high probability paths of length t between two points. A related idea has been recently introduced in (Giannuzzi et al. 2016) based on Independent Component Analysis (ICA).

3.4.1.3 Sparse Diffusion Maps Representation

An acute problem arises from the need to compute singular vectors of a large symmetric matric. To reduce the computational complexity, a truncated similarity matrix can be defined by thresholding matrix M, as

$$\hat{M} = \begin{cases} 0 & \text{if } i = j \\ m_{ij} & \text{if } m_{ij} \geq th. \\ 0 & \text{if } m_{ij} \leq th \end{cases}$$

A drawback of DMs and other nonlinear dimensionality reduction methods is the absence of an explicit mapping between the input variables and the variables parametrizing the low dimensional embeddings. This can make it challenging to identify a clear interpretation of the low-dimensional collective modes. Further, conventional data mining methods are designed to deal with static data sets where the ordering of records has nothing to do with the patterns of interest.

3.4.1.4 Time-Domain Decompositions

As pointed out in Arvizu and Messina (2016), a limitation of the DM framework (in the context of power system monitoring) is that it makes no use of the temporal patterns of the original measurement matrix which makes its comparison to other analytical approaches difficult. In the development that follows, the diffusion coordinates are given a dynamic context by expressing the data matrix in terms of the local coordinates.

The key result is that the snapshot matrix in (3.3) can be approximated by a linear combination of the first non-trivial coordinates by projecting the data into a dynamically meaningful (slow) subspace, such that

$$\widehat{X} = \begin{bmatrix} \hat{x}_1 & \hat{x}_2 \dots & \hat{x}_d \end{bmatrix} = \sum_{j=0}^{m} a_j \psi_j^T = \underbrace{\sum_{j=0}^{d} a_j(t)\psi_j^T}_{\text{Relevant system behavior}} + \underbrace{\sum_{k=d+1}^{m} a_k(t)\psi_k^T}_{\text{non-essential coordinates}},$$

(3.16)

where the $\hat{x}_j(t) = [\hat{x}_j(t_1) \quad x_j(t_2) \quad \cdots \quad x_j(t_N)]^T, j = 1, \dots, d$ are the time coefficients of the diffusion coordinates, and the superposition of the terms $a_j \psi_j^T$ approximates the entire data sequence, where the first few temporal components represent dominant temporal patterns that capture the dominant system behavior.

In the same spirit as with principal component analysis shown in Hannachi et al. (2007), the time coefficients are found by projecting the snapshot trajectories onto the span of the first d eigenfunctions as

$$a_j = X\psi_j, \quad j = 1, \dots, d.$$

A trajectory can be projected onto a subset of selected eigenvectors so only motion along the selected vectors is allowed. It should be stressed that the temporal coefficients are not harmonically pure modes.

3.4.1.5 Physical Interpretation of Diffusion Coordinates

The time coefficient $a_j(t)$ provides a good approximation to the collective motion with the slowest time scale. Suppose the temporal coefficients in (3.16) can be approximated as $a_j(t) = A_j(t)\cos(\omega_j t + \varphi_j)$, $j = 1, \dots, d$, where $A_j(t)$ is the instantaneous amplitude and ω_j are the mode's frequencies. The temporal vector, a_j, can then be defined as

$$a_j = \begin{bmatrix} a_j(t_o) & a_j(t_1)\cdots & a_j(t_N) \end{bmatrix}^T.$$

Frequency and amplitude estimation can then be obtained, noting that (for a lossless system) the total energy $PE = KE$ is conserved, namely

$$\frac{1}{2}\omega_i^2 \sum_{t=1}^{N} x(t)_i^2 - \frac{1}{2}\sum_{t=1}^{N} \dot{x}(t)_i^2 = 0, \tag{3.17}$$

where PE and KE denote potential and kinetic energy (Guo et al. 2011).

It follows that the ith frequency can be expressed as

$$\omega_i^2 = \frac{\sum_{t=1}^{N} \dot{x}(t)_i^2}{\sum_{t=1}^{N} x(t)_i^2}. \tag{3.18}$$

The extension to the multivariate case is immediate. Assuming orthogonality, $a_j^T a_j = 1$, it can be proved that

$$\omega_j \approx \sqrt{\dot{a}_j^T \dot{a}_j / a_j^T a_j}. \tag{3.19}$$

Unlike similar formulations such as principal component analysis, the time coefficients are not necessarily zero mean functions and exhibit a degree of correlation among each other. A quantitative measure of the correlation between two modal components is given by $\varepsilon_{ij} = a_i^T(t)a_j(t) / \|a_i(t)\| \|a_j(t)\|$, where $\varepsilon_{ij} = 0$ if $a_i(t) \neq a_j(t)$, $\varepsilon_{ij} = 1$ if $a_i(t) = a_j(t)$. Also, energy exchange between two-time coordinates or modes can be obtained. Defining $E_k = a_k^T(t)a_k(t)$, $k = 1, \dots, d$, energy exchange is obtained if $dE_i / dt + dE_j / dt = 0$.

The following properties can be derived from this model:

1. For simple geometries, a physical interpretation is readily available; the diffusion coordinates $\psi_j(x)$ give the observability of each mode in the states, with $j = 1$ being the dominant mode.

2. In the more general case, however, the vector can be associated with a given frequency range as $\psi_j(x) = \begin{bmatrix} \psi_{1j}(x_1) & \psi_{2j}(x_2)\dots & \psi_{mj}(x_m) \end{bmatrix}^T$, where ψ_{kj} denotes the kth component of the coordinate vector associated with the observable x_k.

As a generalization of the above approach, one may define other metrics. A global modal index (GMI) associated with a given observable, $\hat{x}_k$, was defined in Arvizu and Messina (2016) as:

$$GMI(\hat{x}_k) = \left[\frac{1}{\sigma_1}\psi_1(\hat{x}_k) \quad \frac{1}{\sigma_1}\psi_1(\hat{x}_k)... \quad \frac{1}{\sigma_1}\psi_1(\hat{x}_k)\right]^T. \qquad (3.20)$$

The above approach offers two major advantages over linear dimensionality reduction methods: diffusion maps are nonlinear, and they preserve local geometries.

Example 3.1 discusses the application of this approach to simulated data.

Example 3.1

As an illustration of the application of this method, consider actual measured data collected by 6 PMUs. Figure 3.6 shows time histories of the recorded measurements.

For the present analysis, the data set can be expressed in the form

$$X_f = \begin{bmatrix} f_1 & f_2... & f_6 \end{bmatrix} \in \Re^{1402 \times 6},$$

where $f_k = \begin{bmatrix} f_{PMU_k}(t_o) & f_{PMU_2}(t_k)... & f_{PMU_k}(t_N) \end{bmatrix}^T$, $k = 1,...,6$ is a column vector of frequency measurements, and 1402 snapshots are considered.

From the development above, the data matrix is expressed in the form

$$\hat{X} = \sum_{j=o}^{m} a_j \psi_j^T.$$

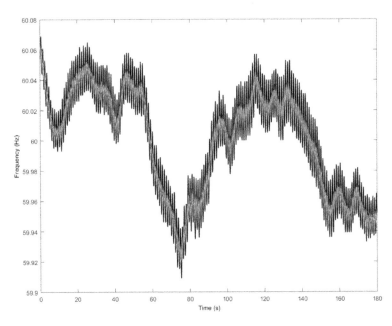

FIGURE 3.6
Frequency data collected by PMUs.

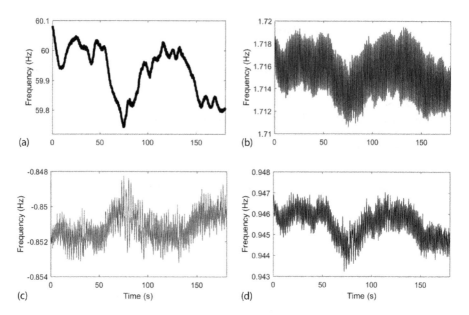

FIGURE 3.7
Temporal coefficients, (a–d) a_o, a_1, a_2, and a_3.

Figure 3.7 show the extracted temporal coefficients. Inspection of the temporal coefficients in Figure 3.7, shows that a_o captures the overall trend, while a_1 is seen to capture the main oscillatory behavior. The other components are found to have a rather negligible contribution to the data.

3.4.2 The Meila–Shi Algorithm

A closely related algorithm to diffusion maps is the Meila–Shi algorithm introduced in Meila and Shi (2001). Given a real $m \times N$ data matrix $\mathbf{X}$, the Meila–Shi algorithm defines the similarity matrix $\mathbf{S}$, such that

$$S = \left[s_{ij} \right] = XX^T,\tag{3.21}$$

where S is a square, symmetric matrix (Scott et al. 2017). The adjacency matrix, is

$$A = \left[a \right] = \frac{s_{ij}}{\sqrt{\sum_{k=1} s_{ik} \sum_{k=1} s_{jk}}}.\tag{3.22}$$

The row-stochastic matrix A (a Markov or transition matrix) is then defined as

$$P = [p_{ij}] = \frac{S_{ij}}{\sum_{k=1} S_{ik}}. \tag{3.23}$$

Similar to the diffusion map case, it can be easily verified that matrices P and A are related to each other by a similarity transformation.

Define now the diagonal matrices

$$R = [r_{ij}] = \begin{bmatrix} \sum_{j=1}^{m} S_{1j} & 0... & 0 \\ 0 & \sum_{j=1}^{m} S_{2j} & 0 \\ 0 & 0... & \sum_{j=1}^{m} S_{mj} \end{bmatrix}, \text{ and }$$

$$D = [d_{ij}] = \begin{bmatrix} \sqrt{\dfrac{\sum_{j=1}^{m} S_{1j}}{\sum_{j=1}^{m} S_{1j} \sum_{j=1}^{m} S_{1j}}} & 0... & 0 \\ 0 & \sqrt{\dfrac{\sum_{j=1}^{m} S_{1j}}{\sum_{j=1}^{m} S_{1j} \sum_{j=1}^{m} S_{1j}}} & 0 \\ 0 & 0... & \sqrt{\dfrac{\sum_{j=1}^{m} S_{1j}}{\sum_{j=1}^{m} S_{1j} \sum_{j=1}^{m} S_{1j}}} \end{bmatrix}.$$

Mathematically, matrices S and P are related by the similarity transformation $S = RP$. It then follows that

$$A = DR^{-1}SD^{-1} = DPD^{-1}. \tag{3.24}$$

Clearly, matrices A and P share the same eigenvalues. Let $\lambda_o = 1$, and $\lambda_o > \lambda_i > ... > \lambda_{i+1}$ and $\psi_i, \phi_i, i = 1,2,3...$ be the eigenvalues and eigenvectors of A and P, respectively. It readily follows that their eigenvectors are related to each other by the transformation

$$\psi_i = D^{-1}\phi_i.$$

A similar algorithm to that of the diffusion maps algorithm can obtained for extracting spectral properties. Details are omitted.

3.4.3 Distributed Stochastic Neighbor Embedding (t-SNE) Method

The stochastic neighbor embedding method converts the original (high-dimensional) Euclidean distances into conditional probabilities (the occurrence of one event depends on the occurrence of another) that represent similarities (Van der Matten and Hinton 2008).

Given two dynamic trajectories $x_i(t), x_j(t)$, the probability density function of the neighboring data points for $x_i(t)$ is given by

$$P_{j|i} = \frac{\dfrac{\exp\left(\|x_i - x_j\|\right)}{\sigma}}{\displaystyle\sum_{k \neq i} \dfrac{\exp\left(\|x_i - x_j\|\right)}{\sigma}}, \tag{3.25}$$

where σ is the variance of the Gaussian that is centered on x_i, $d_{ij}^2 = \|x_i - x_2\|^2$, and $P_{j|i}$ represents conditional probability.

By construction, the conditional probability is non-symmetrical ($p_{ij} \neq p_{ji}$) and has a similar interpretation to the Markov transition matrix in (3.11). Following Van der Maaten and Hinton (2008), the similarity of data points is calculated as the joint probability

$$p_{ij} = \frac{P_{j|i} + P_{i|j}}{2N}.$$

A similar joint probability describing similarity can be computed in the low dimensional space.

Drawing on the diffusion-based approach, the probability matrix is defined as $P_{ij} = [p_{ij}]$, where $p_{ii} = 0$. In a variation of the above approach, the variance is calculated using the procedure in Hannachi et al. (2007), and a spectral decomposition is applied to the Markov transition matrix.

The following sections extend these ideas to the multisensory case and discuss their application to clustering power system data.

3.4.4 Multiview Diffusion Maps

The analysis of heterogeneous data coming from multiple sensors (multiview data) needs to be incorporated into current data mining and data fusion techniques. Conceptually, a lower dimensional representation that preserves the interactions between multidimensional data points is sought.

A key issue is the integration of multimodal data involving a wide range of dimensionalities across the different types or modalities. A possible solution to overcome these representational differences is to first project the data streams into a space where the scale and dimensionality differences are removed; this is referred to as a *meta-space*.

Following Leidendaum et al. (2015), consider a set of multiple observations, $X^l, l = 1, \ldots, L$, where each set is a high-dimensional data set $X^l = \begin{bmatrix} x_1^l & x_2^l \ldots & x_L^l \end{bmatrix} \in \mathfrak{R}^{N \times m}$. For ease of exposition, it is assumed that all measurements correspond to the same sensors, and that all records are of the same length, N. This is the case, for instance, for data from different tests or perturbations recorded by the same set of sensors. Generalizations to this basic approach are discussed in Chapter 5.

Figure 3.8 illustrates the nature of the multiview scenario used in the adopted model. For each data set, the multiview diffusion Gaussian kernel can be defined as

$$\left[K^l \right] = \left[k_{ij}^l \right] = \exp\left(-\frac{\left\| x_i^l(t) - x_j^l(t) \right\|}{\varepsilon_i \varepsilon_j} \right), l = 1, \ldots, L, \tag{3.26}$$

where the $\varepsilon_i \varepsilon_j$ represent local scaling associated with data set X^l.

Inspired by other multiview integration techniques, the symmetric positive-definite multiview kernel can be defined as

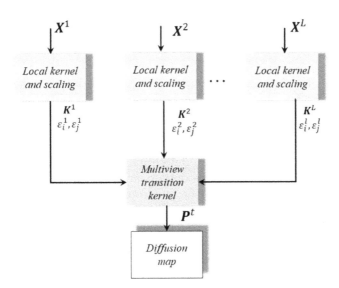

FIGURE 3.8
Data sets for the adopted Multiview diffusion framework.

TABLE 3.3

Nonlinear Model Reduction Techniques

Method	References
Local linear embedding (LLE)	Roweis and Saul (2000)
Laplacian eigenmaps	Belkin and Niyogi (2003)
Hessian eigenmaps	Donoho and Grimes (2003)
Isomap	Tenenbaum et al. (2000)
Logmaps	Brun et al. (2005)

$$\widehat{K} = \begin{bmatrix} 0_{M \times M} & K^1 K^2 & K^1 K^3 & \cdots & K^1 K^p \\ K^2 K^1 & 0_{M \times M} & K^2 K^3 & \cdots & K^2 K^p \\ K^3 K^1 & K^3 K^2 & 0_{M \times M} & \cdots & K^3 K^p \\ \vdots & \vdots & & \ddots & \vdots \\ K^p K^1 & K^p K^2 & K^p K^3 & \cdots & 0_{M \times M} \end{bmatrix} \in \Re^{l \times m \times l \times m}. \tag{3.27}$$

Like the univariate case, matrix $\widehat{K}$ is symmetric and positive definite and has a spectral decomposition of interest. Similar to the single type case, these kernels $K^1, \ldots, K^l$ are normalized and then the spectral decomposition can be obtained. This representation is not unique, and several enhancements are possible borrowing ideas from other multiview data integration techniques. An algorithm for this approach is given in Chapter 5.

Other sparse spectral techniques can also be applied to reduce model dimensionality. Table 3.3 lists some prevalent nonlinear dimensionality reduction methods—see Strange and Zwiggelaar (2014) for details.

3.5 Clustering Schemes

Upon determining a low-dimensional representation, trajectory clustering techniques can be applied to determine coherent behavior of selected variables as well as to determine outliers and pattern mining (David and Averbuch 2012).

Traditionally, linear analysis techniques including Principal Component Analysis (PCA) and related techniques have been used to identify clusters from measured data. A limitation of these approaches, however, is their inability to capture nonlinearities.

To describe these approaches, it is useful to introduce the notion of metrics. Assume that the system trajectories from Figure 3.1 have been clustered into k classes or clusters exhibiting common behavior. Associated with each cluster is a *centroid* representing the mean fluctuation.

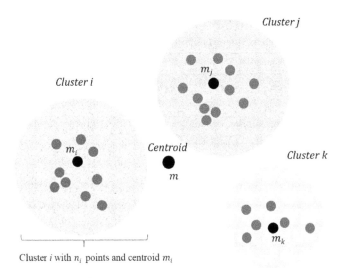

Cluster i with n_i points and centroid m_i

FIGURE 3.9
Clusters illustrating the notion of local centroids and global mean. Filled circles represent points or sensors, and m denotes the average of all centroids in the data set.

Referring to Figure 3.9, the mean value of each cluster (centroid) is calculated as the mean of all the data points belonging to that cluster, namely

$$m(c_j) = \frac{1}{n_j} \sum_{x_i \in c_i}^{n_j} (x_i), \tag{3.28}$$

where n_j is the number of data points belonging to cluster j, and x_i is a data point. Associated with this measure, the *cluster dispersion* can be defined as

$$D(c_j) = \frac{1}{n_j} \sum_{x_i \in c_j}^{n_j} \underbrace{(x - c_j)}_{\text{distance function}}$$

and is illustrated in Figure 3.9.

Several definitions of interest associated with the notion of inter-cluster distances are summarized in Table 3.4.

In the remainder of this section, two clustering techniques are examined: *K*-means clustering and Fuzzy clustering.

3.5.1 *K*-Means Clustering

One of the most fundamental approaches to data clustering is the K algorithm. *K*-means clustering has been extensively applied to cluster power

TABLE 3.4

Inter-Cluster Distance Measures

Measure	Approximation
Dispersion	$D(c_j) = \dfrac{1}{n_j} \sum_{x_i \in c_j} (x - c_j)$
Intra-cluster distance	$D(c_i) = \dfrac{1}{n} \sum_{x \in c_i} (x - C_i)$

system data (Dutta and Overbye 2014; Arvizu and Messina 2016). The application of K-means clustering with DMs is discussed in (Arvizu and Messina 2016).

In the standard k-center problem, the objective is to compute a set of k center points to minimize the maximum distance from any point S to its closest center, or equivalently, the smallest radius such that S can be covered by k disks of this radius (Zelnik-Manor and Perona 2003). In the discrete k-center problem, the disk centers are drawn from the points of S, and in the absolute k-center problem the disk centers are unrestricted.

This can be posed as the solution of the optimization problem

$$f = \sum_{k=1}^{K} \sum_{x_i \in c_i}^{n} \left\| x_i - m_k \right\|^2 , \tag{3.29}$$

where K is the number of clusters, c_i is the ith cluster, x_i is a data point, and m_k is the centroid of the kth cluster.

3.5.2 Dispersion

Dispersion is a measure of average distance between members of the same cluster (Laffon and Lee 2006). With reference to Figure 3.2, for a microstate segmentation with k clusters, the dispersion D_k is calculated as the sum of squares between the members of each microstate cluster.

Consider a cluster C_i with n points. The intra-cluster dispersion, D_k, can be defined as

$$D_k = \sum_{i=1}^{n} \sum_{j=1}^{n} \left\| x_i(t) - x_j(t) \right\|^2. \tag{3.30}$$

An outline of the K-means algorithm is as follows:

1. Select the number of clusters K. Select K points as the initial centroids; choose m, such that $2 \le K \le n$.
2. Assign each point to the nearest centroid.

3. Recompute the coordinates of the centroids.
4. Repeat steps 2 and 3 until the centroids no longer change position.

Criteria to automatically determine the number of clusters are given in Zelnik-Manor and Perona (2003).

The algorithm is simple to implement, and computational requirements are small. Key advantages of these algorithms are their ability to deal with large-sized problems and computational speed. Its main disadvantage is that each different initial assignments of centroids converges to a different set of clusters. However, the performance of the method is poor in high dimensions.

Extensions to these ideas are discussed next within the framework of Fuzzy cluster algorithms.

3.5.3 C-Means Clustering

Fuzzy c-means clustering (FCM) is an exploratory data-analysis method that identifies groups of samples with similar compositions. Formally, fuzzy c-means clustering can be posed as the solution of the energy function:

$$f(U,v) = \sum_{k=1}^{K} \sum_{i=1}^{n} u_{ki}^{m} \|x_i - m_k\|^2 ,$$ (3.31)

where m_k are the cluster centers, the u_{ki}^{m} represent associated membership likelihoods (the degree of membership), $0 \le u_{ki} \le 1$, of the trajectory x_i being associated with the cluster center C_k.

The optimization problem is subject to the constraints

$$\sum_{k=1}^{K} u_{ki} = 1, i = 1, \dots, n$$

$$u_{ki} \ge 0, k = 1, \dots, K, i = 1, \dots, n.$$

Formally, given n dynamic trajectories, the problem can be stated as follows:

$$f() = \sum_{k=1}^{K} \sum_{i=1}^{n} u_{ki}^{m} \|x_i - C_k\|^2 , \text{ subject} t$$

$$\sum_{k=1}^{K} u_{ki} = 1, i = 1, \dots, n$$ (3.32)

$$u_{ki} \ge 0, k = 1, \dots, K, i = 1, \dots, n.$$

The following algorithm, Fuzzy c-means Algorithm, is adopted for computation of clusters.

Fuzzy c-means Algorithm

---

1. Initialize membership values u_{ki}.
2. Calculate the K fuzzy cluster centers using

$$C_k = \frac{\sum_{k=1}^{K} u_{ki} x_i}{\sum_{i=1}^{n} u_{ki}^m}, k = 1, \ldots, K.$$

3. Update membership values using

$$u_{k,i} = \frac{1/\|x_i - C_k\|^{2/(m-1)}}{\sum_{j=1}^{K} \left(1/\|x_i - C_j\|^{2/(m-1)}\right)}, k = 1, \ldots, K, i = 1, \ldots, n.$$

4. Evaluate the energy function (3.32). If the improvement in the objective is below a threshold, go to step 5; otherwise go to step 2.
5. Output cluster centers $C_k \in \Re^{d(T+1)}, k = 1, \ldots, K$ and membership likelihoods $u_{ki} \in [0,1], k = 1, \ldots, K, i = 1, \ldots, n.$

The algorithm is simple to implement, fast, and computationally-efficient for large numbers of trajectories.

Once the number of clusters is defined, c-means clustering can be used to determine the time evolution of the centroids, as well as to measure fluctuations with respect to the mean value or deviations between clusters.

Example 3.2

As a practical example, the application of clustering techniques is demonstrated on simulated data from a 14-generator, 59-bus, and 5-SVC system adapted from a benchmark model published by IEEE Task Force on Benchmark Systems for Stability Control (2017). A diagram of this system is shown in Figure 3.10.

The system dynamic behavior is characterized by four weakly connected regions resulting in several local and inter area electromechanical modes. The principal system behavior is characterized by four main inter-area modes at about 0.289, 0.381, 0.444, and 0.579 Hz. The contingency of interest is a 1% MW load switching at bus 217. This contingency is found to excite a major inter-area mode at about 0.45 Hz—refer to Table 3.5. Figure 3.11 shows the speed deviations of the system generators.

Prony analysis of the temporal coefficients in Table 3.6 shows the accuracy of the low-dimensional model. The temporal coefficient, a_o in Figure 3.12 is seen to capture the overall trend; coefficients a_1 and a_2, on the other hand, are seen to capture the main oscillatory behavior.

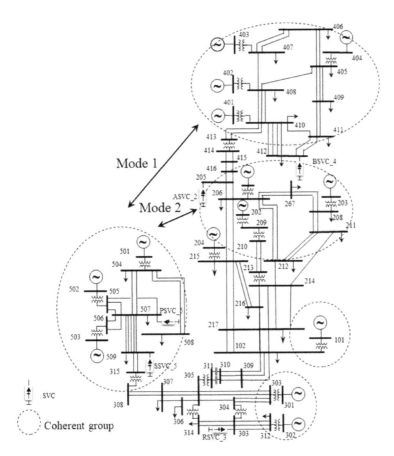

FIGURE 3.10
Single-line diagram of the 14-machine Australian test system.

TABLE 3.5

The Slow Eigenvalues of the System

Mode	Eigenvalue	Frequency (Hz)	Damping (%)
1	$-0.5278 \pm j1.7965$	0.2859	28.19
2	$-0.3809 \pm j2.3994$	0.3819	15.74
3	$-1.4341 \pm j2.8063$	0.4446	15.68
4	$-1.0432 \pm j3.6407$	0.5794	27.54

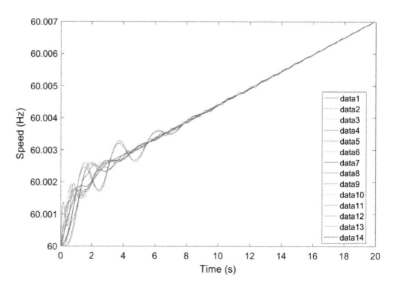

FIGURE 3.11
Speed deviations of system generators.

TABLE 3.6

Prony Analysis Results of the Temporal Modes in Figure 3.12

Coefficient	Frequency (Hz)	Damping (%)
a_o	0.4858	15.30
a_1	0.4591	13.16

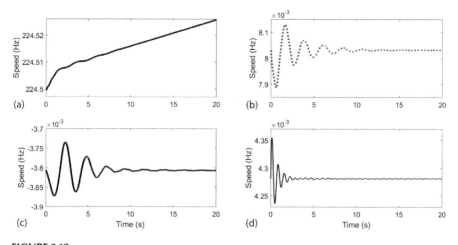

FIGURE 3.12
Temporal coefficients in (3.16). (a) $a_o(t)$ (b) $a_1(t)$, (c) $a_2(t)$, and (d) $a_3(t)$.

3.5.3.1 Trajectory Clustering: Diffusion Maps Analysis

Application of DMs results in two main spatiotemporal modes ($d = 2$). Insight into the nature of spatial behavior can be gleaned from Figure 3.13 that depicts the spatial modes ψ_j, $j = 1, \ldots, d$, $d = 4$. Figure 3.14a shows the spectrum of the Markov matrix along with the extracted clusters using DMs. A significant gap between the first and the remaining eigenvalues is evident suggesting that a single mode dominates dynamic behavior.

Using DM, the rotor angle data was clustered into 5 clusters. As shown in Figure 3.14b, the two leading eigenvectors are sufficient to provide a clustering perspective of coherent behavior; results are consistent with expected coherent behavior in Figure 3.10.

3.5.3.2 C-Means Clustering

To further analyze the applicability of clustering techniques to multimodal data, voltage and bus frequency signals were analyzed simultaneously. Analysis of c-means clustering in Figures 3.15 and 3.16 illustrates the time evolution of each coherent group and their associated centroids. As shown, this approach allows to further decompose the initial c clusters into subclusters depicting similar behavior.

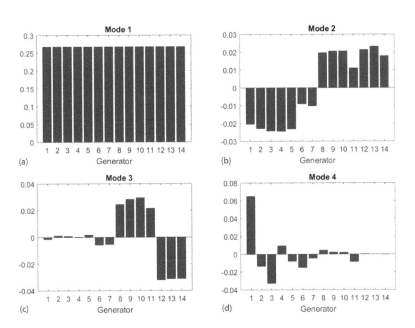

FIGURE 3.13
First few eigenvectors extracted using DMs. (a) Mode 1, (b) Mode 2, (c) Mode 3, and (d) Mode 4.

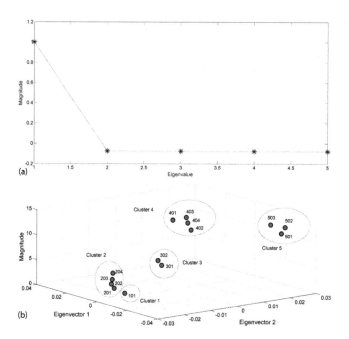

FIGURE 3.14

Clustering of speed deviation data. (a) Eigen spectrum of matrix M. (b) Cluster of rotor angle deviations.

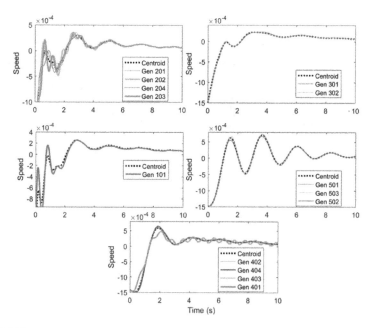

FIGURE 3.15

Time evolution of speed deviation clusters. Speed deviations.

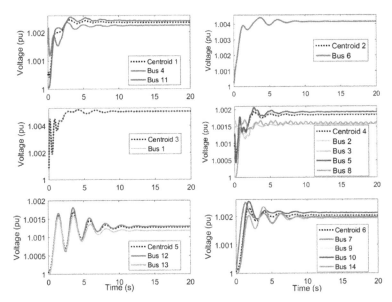

FIGURE 3.16
Time evolution of speed deviation clusters. Bus terminal voltages.

3.6 Detrending and Denoising of Power System Oscillations

Detrending and denoising is a first step before meaningful results can be obtained from mining. Experience with the analysis of complex data shows that data mining and data fusion results can be biased by the trend components. Measured data will always contain exhibit trends and other irregular components that can prevent or complicate analysis and extraction of special features of interest in system behavior.

Among the common methods for extracting trends, there are several standard approaches including mean average techniques, wavelet shrinkage, and other heuristic approaches (Messina et al. 2009). A summary of some recent approaches to power system data mining and denoising is shown in Table 3.7.

TABLE 3.7

Detrending (denoising) Techniques

Method	References
Wavelet shrinkage	Baxter and Upton (2002)
Hilbert-Huang analysis	Huang et al. (1998)
Kalman filter, Moving average, Heuristic approaches	Messina et al. (2009)

References

Arvizu, C. M., Messina, A. R., Dimensionality reduction in transient simulations: A diffusion maps approach, *IEEE Transactions on Power Delivery*, 31(5), 2379–2389, 2016.

Barocio, E., Pal, B. C., Thornhill, N. F., Messina. A. R., A dynamic mode decomposition framework for global power system oscillation analysis, *IEEE Transactions on Power Systems*, 30(6), 2902–2912, 2015.

Baxter, P. D., Upton, G. J. G., Denoising radio communication signals by using iterative wavelet shrinkage, *Applied Statistics*, 51(4), 393–403, 2002.

Belkin, M., Niyogi, P., Laplacian eigenmaps for dimensionality reduction and data representation, *Neural Computation*, 15, 1373–1396, 2003.

Bezdek, J. C., Ehrlich, R., Full, W., FCM: The Fuzzy c-means clustering algorithm, *Computers and Geosciences*, 10(2–3), 191–203, 1984.

Bezdek, J. C., *Pattern Recognition with Fuzzy Objective Function Algorithms*, Plenum Press, New York, 1981.

Borcard, D., Legendre, P., Drapeau, P., Partialling out the spatial component of ecological variation, *Ecology*, 73, 1045–1055, 1992.

Bro, R., Smilde, A. K., Centering and scaling in component analysis, *Journal of Chemometrics*, 17, 16–33, 2003.

Brun, A., Westin, C. F., Herberthson, M., Knutsson, H., Fast Manifold learning based on Riemannian normal coordinates, *Lectures Notes in Computer Science*, 3540, 920–929, 2005.

David, G., Averbuch, A., Hierarchical data organization, clustering and denoising via localized diffusion folders, *Applied and Computational Harmonic Analysis*, 33(1), 1–23, 2012.

Donoho, D. L., Grimes, C., Hessian eigenmaps: Locally linear embedding techniques for high-dimensional data, *Proceedings of the National Academy of Sciences of the United States of America*, 100(10), 5591–5596, 2003.

Dutta, S., Overbye, T. J., Feature extraction and visualization of power system transient stability results, *IEEE Transactions on Power Systems*, 29(2), 966–973, 2014.

Giannuzzi, G., Pisani, C., Sattinger, W., Generator coherency analysis in ENTSO-E continental system: Current status and ongoing developments in Italian and Swiss case, *IFAC-PapersOnLine*, 49(17), 400–406, 2016.

Griffith, D. Eigenfunction properties and approximations of selected incidence matrices employed in spatial, *Linear Algebra and Its Applications*, 321, 95–112, 2000.

Guo, X., Energy-weighted modes selection in reduced-order nonlinear simulations, *52nd AIAA/ASME/ASCE/AHS/ASC Structures, Structural Dynamics and Materials Conference*, AIAA 2011-2063, Denver, CO, 2011.

Hannachi, A., Jolliffe, I. I., Stephenson, D. B., Empirical orthogonal functions and related techniques in atmospheric science: A review, *International Journal of Climatology*, 27, 1119–1152, 2007.

Huang, N. E, Shen, Z., Long, S. R., Wu, M. C., Shih, H. H., Zheng, Q., Yen, N., Tung, C. C., Liu, H. H., The empirical mode decomposition and the Hilbert spectrum for nonlinear and non-stationary time series analysis, *Proceedings of the Royal Society*, 454(1971), 903–995, 1998.

IEEE Task Force on Benchmark Systems for Stability Control, Benchmark models for the analysis and control of small-signal oscillatory dynamics in power systems, *IEEE Transactions on Power Systems*, 32(1), 715–722, 2017.

Kezunovic, M., Abur, A., Merging the temporal and spatial aspects of data and information for improved power system monitoring applications, *Proceedings of the IEEE*, 93(11), 1909–1919, 2005.

Lafon, S., Lee, A. B., Diffusion maps and coarse graining: A unified framework for dimensionality reduction, graph partitioning, and data set parameterization, *IEEE Transactions on Pattern Analysis and Machine Intelligence*, 28(9), 1393–1403, 2006.

Lindenbaum, O., Yeredor, A., Salhov, M., Averbuch, A., Multiview diffusion maps, 24(2), 451–505, 2015.

Meila, M., Shi, J., Learning segmentation by random walks, *Neural Information Processing Systems*, 13, 873–879, 2001.

Messina, A. R., *Wide-Area Monitoring of Interconnected Power Systems*, IET, Power and Energy Series, 77, Stevenage, UK, 2015.

Messina, A. R., Vittal, V., Extraction of dynamic patterns from wide-area measurements using empirical orthogonal functions, *IEEE Transactions on Power Systems*, 22(2), 682–692, 2007.

Messina, A. R., Vittal, V., Heydt, T. G., Browne, T. J., Nonstationary approaches to trend identification and denoising of measured power system oscillations, *IEEE Transactions on Power Systems*, 24(4), 1798–1807, 2009.

Newman, M., *Networks*, 2nd ed., Oxford University Press, Oxford, UK, 2018.

Rovnyak, S. M., Mei, K., Dynamic event detection and location using wide area phasor measurements, *European Transactions on Electrical Power*, 21(4), 1589–1599, 2011.

Roweis, S. L., Saul, L. K., Nonlinear dimensionality reduction by locally linear embedding, *Science*, 290(22), 2323–2326, 2000.

Scott, T. C., Therani, M., Wang, X. M., Data clustering with quantum mechanics, MDPI, *Mathematics*, 5(1), 5, 2017.

Strange, H., Zwiggelaar, R., *Open Problems in Spectral Dimensionality Reduction*, Springer Briefs in Computer Science, New York, 2014.

Taylor, K. M., Procopio, M. J., Young, C. J., Meyer, F. G., Estimation of arrival times from seismic waves: A manifold-based approach, *Geophysical Journal International*, 185(1), 435–452, April 2011.

Tenenbaum, J. B., De Silva, B., Langford, J. C., A global geometric framework for nonlinear dimensionality reduction, *Science*, 290(5500), 2319–2323, 2000.

Van den Berg, R., Hoefsloot, H. C. J., Smilde, A. K., Van der Werf, M. J., Centering, scaling, and transformations: Improving the biological information content of metabolomics data, *BMC Genomics*, 7(1), 142, 2006.

Van der Maaten, L., Hinton, G., Visualizing data using t-SNE, *Journal of Machine Learning Research*, 9, 2579–2605, 2008.

Young, P. C., Pedregal, J., Tych, W., Dynamic harmonic regression, *Journal of Forecasting*, 18, 369–394, 1999.

Zavala, A. J., Messina. A. R., A dynamic harmonic regression approach to power system modal identification and prediction, *Electric Power Components and Systems*, 42(13), 1474–1483, 2014.

Zelnik-Manor, L., Perona, P., Self-tuning spectral clustering, *Proceedings of the 17th International Conference on Neural Information Processing Systems*, 1601–1608, Vancouver, Canada, December 2003.

4

Spatiotemporal Data Mining

4.1 Introduction

Power system data is inherently spatiotemporal, noisy, heterogeneous, and uncertain. Sources of uncertainty and variability include natural variability associated with random load fluctuations (Trudnowski and Pierre 2009), forced oscillation (Sarmani and Venkatasubramanian 2016), and equipment malfunction, as well as spatiotemporal evolutionary processes associated with load and topology changes and control actions, among other issues. In addition, data may be available at various temporal and spatial resolutions from various sensing technologies such as PMU, dynamic frequency recorders, and SCADA systems.

Data from dynamic sensors may be highly correlated, especially when sensors are located close to each other. Additionally, power system phenomena operate and interact on multiple spatiotemporal scales. These factors may result in numerical issues and increase the dimension of the state models. Further, because of interactions between dynamic devices caused by synchronizing forces, the true dimensionality of the system is far lower than that of the original space and the resultant data measurements are often sparse. As a result, some areas may be underrepresented while others may be overrepresented (Messina 2009).

The complexity of spatiotemporal data and intrinsic relationships may limit the usefulness of conventional data science techniques for extracting spatiotemporal patterns. It is therefore necessary to develop new techniques that efficiently summarize and discover trends in spatiotemporal data in order to facilitate decision making. To be of practical use, data mining techniques should consider space and time in an integrated manner.

As the use of measurement sensors becomes more extended and the recorded data become larger and more prevalent, more sophisticated methods to quickly and accurately mine larger numbers of trajectories for relevant information will have to be developed.

Emerging areas of research in time-series data mining include, among others (Cheng et al. 2013; Ghamisi et al. 2019; Shekhar et al. 2015):

- Spatiotemporal prediction and forecasting,
- Spatiotemporal sequential pattern mining,
- Spatiotemporal association rule mining,
- Spatiotemporal clustering and classification, and
- Spatiotemporal visualization.

This chapter introduces the notion of spatiotemporal data mining. Emphasis is focused on the simultaneous spatial and temporal study of patterns over time. Other aspects of spatiotemporal data mining are discussed later in this book. Applications of several spatiotemporal data mining techniques to track the evolving dynamics of power system modes in both simulated and measured data are presented. The issues of computational efficiency and memory requirements are also discussed.

4.2 Data Mining and Knowledge Discovery

Spatiotemporal data mining can be broadly defined as the extraction of unknown and implicit knowledge, structures, relationships, or patterns from data collected at two or more locations (Cheng et al. 2013). These relationships are referred to as *patterns* (Hand et al. 2001) and they often have a physical interpretation in terms of the fundamental modes of motion in small signal stability analysis (Messina and Vittal 2007). In this sense, *knowledge discovery* describes the process of extracting useful and meaningful information from data, though sometimes the terms *data mining* and *knowledge discovery* are used interchangeably (Chu 2014).

Knowledge discovery incorporates a number of processes and algorithms from the preparation of raw data prior to the application of data mining to visualization of the final result. Visualization is also an integral part of this process. The processes are often problem (application) dependent and may include data selection and preprocessing, data transformation, data mining, and data interpretation and evaluation. A survey on time-series data mining is presented in (Esling and Agon 2012).

Figure 4.1 illustrates the process of knowledge discovery from measured data. Four main steps are of interest here (Chu 2014; Fayyad et al. 1996)

1. Data cleansing and applying filters for outliers,
2. Data integration and data selection,
3. Data transformation, and
4. Data mining and pattern evaluation.

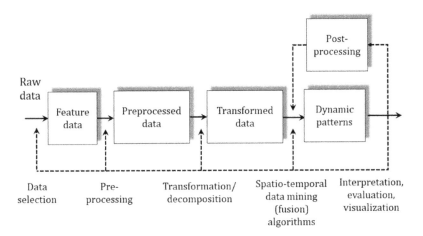

FIGURE 4.1
Basic data mining and knowledge discovery process pipeline. (Adapted from Fayyad, U. et al., *AI Mag.*, 17, 37–53, 1996.)

Data cleansing incorporates techniques for handling missing data, noise correction, and artifact removal. Through preprocessing techniques, an accurate representation of measured signal where noise and other artifacts are removed can be obtained. For some applications, data (feature) selection is needed to compute features of interest or focus on specific behavior (i.e., power oscillation monitoring).

In some applications, data transformation is used to reduce system dimensionality. Methods such as discrete Fourier transform (DFT), wavelet decomposition, or time domain decomposition can be used to characterize specific behaviors of interest. Such approaches are useful for examining oscillatory behavior and/or forecasting future system performance (Liu et al. 2007; Farantatos and Amidan 2019). Automated data mining and pattern evaluation techniques can also be utilized to extract information of interest.

Open issues include the selection of the appropriate scale of analysis, handling multi-scale data, clustering with irregular background clutter, and finding automatically the number of clusters.

4.3 Spatiotemporal Modeling of Dynamic Processes

This section provides background on conventional methods that are used for mining data from observational data. The theoretical bases underpinning these methods are derived and equivalences between the several models are developed.

4.3.1 Conceptual Framework

Spatiotemporal time series can be seen as sequences of values or events changing with time. A typical time series contains a trend, an oscillatory component, and an irregular component or noise (Harvey 1990).

Following the notation of the previous section, let $X = [x_1 \ x_2 \ \cdots \ x_N] \in \mathfrak{R}^{m \times N}$ represent a set of observed (measured) system data at locations $\{1, 2, ..., m\}$, where N represents time and m denotes the number of sensors or system locations.

At its most generic level, multivariate methods for spatial time series seek a representation of the data by a model of the form

$$x_k(t) = \underbrace{a_o(t)\phi_o(x)}_{\substack{\text{Temporal} \\ \text{mean}}} + \sum_{j=1}^{p} a_j(t)\phi_j(x) + \varepsilon(t), k = 1, ..., m \qquad (4.1)$$

where $a_1, a_2, ..., a_p$ is a set of temporal modes of variation, $\phi_j(x), j = 1, ..., p$ is a vector of dimension m that describes the spatial structures of the modes, and $\varepsilon(t)$ is a vector of residual variability not captured by the p modes (Sieber et al. 2016). The goal is to represent the key content of the data in terms of the smallest number of coordinates, p.

The first term on the right-hand side (rhs) in (4.1) represents the slowly changing temporal mean, while the second denotes fluctuations around the mean value. Depending on the nature of the adopted technique, the basis vectors, $\phi_j(x)$, and temporal coefficients, $a_j(t)$, can be orthogonal, non-orthogonal, linear, or nonlinear.

Cases of these models include principal components analysis (PCA), proper orthogonal decomposition (POD), and Koopman Mode Decomposition (KMD), to mention a few methods. More general descriptions are described in Subsections 4.3.2 through 4.3.6.

4.3.2 Proper Orthogonal Decomposition (POD) Analysis

Statistical methods have the potential to extract dominant features from observational data. Because of its importance and relationship with other methods, this section provides a concise introduction to proper orthogonal decomposition analysis and discusses extensions from it to more complex spatiotemporal representations of interest to those working with power system data. These methods can be used to reduce the number of dimensions in data, find patterns in high-dimensional data, and visualize data of high dimensionality.

The Proper Orthogonal Decomposition (POD) method, also called Empirical Orthogonal Function (EOF) analysis, is an optimal technique of finding a basis that spans an ensemble of observations obtained from measurements, tests, or direct numerical simulation (Kutzbach 1967; Hannachi et al. 2007; Webber et al. 2002). Let

$$
X = \begin{bmatrix} x_1 & \cdots & x_N \end{bmatrix} = \begin{bmatrix} x_1(t_1) & x_1(t_2) & \cdots & x_1(t_N) \\ x_2(t_1) & x_2(t_2) & \cdots & x_2(t_N) \\ \vdots & \vdots & \ddots & \vdots \\ x_m(t_1) & x_m(t_2) & \cdots & x_m(t_N) \end{bmatrix} \in \Re^{m \times N}
$$

denote an $m \times N$ observation matrix whose jth column is the x_j measurement vector.

Following Kutzbach (1967), one wishes to determine which vector ϕ has the highest resemblance to all the observation vectors x_j simultaneously, where resemblance is measured with the squared and normalized inner product.

Averaging across all the measurement vectors, x_j, one seeks to maximize the criterion

$$
\frac{1}{N} \left(\phi^T X \right)^2 \frac{1}{\phi^T \phi} = \frac{1}{N} \frac{1}{\phi^T \phi} \left(\phi^T X \right) \left(\phi^T X \right) \tag{4.2}
$$

This is equivalent to maximizing the quantity $\phi^T R \phi$, subject to the orthonormality condition $\phi^T \phi = 1$, where $R = \left(XX^T \right)/N$ is the $m \times m$ symmetric covariance or correlation matrix of the observations.

The optimal POD basis vectors, can be found by solving the objective function

$$
L = \sum_{j=l+1}^{n} \phi^T XX^T \phi - \sum_{i=l+1}^{n} \sum_{j=l+1}^{m} u_{ij} \left(\phi_i^T \phi_j - \delta_{ij} \right). \tag{4.3}
$$

This results in the eigenvalue problem $R\phi = \lambda \phi$, or collecting all eigenvectors in $R\Phi = \Phi\Lambda$, where $\Phi = \begin{bmatrix} \phi_1 & \phi_2 \ldots & \phi_m \end{bmatrix}$ and $\Lambda = \mathrm{diag} \begin{bmatrix} \lambda_1 & \lambda_2 \ldots & \lambda_m \end{bmatrix}$. See Hannachi et al. (2007) for further details. These modal vectors form a set of orthogonal basic vectors for the space of the data.

Defining $C = \Phi^T X \ \Re^{m \times N}$ one can write

$$
x_j = \sum_{i=1}^{m} a_{ij} \phi_i \tag{4.4}
$$

or

$$
x_1 = \sum_{i=1}^{m} a_{1j} \phi_1
$$

$$
x_2 = \sum_{i=1}^{m} a_{2j} \phi_2
$$

$$
\vdots
$$

$$
x_N = \sum_{i=1}^{m} a_{Nj} \phi_N ,
$$

where the a_{kj} are unknown coefficients to be determined. They represent the projection of the $n, (n = 1, \ldots, N)$ observation vector on the ith eigenvector, ϕ_i.

Collecting terms, one has that

$$X = \Phi A, \tag{4.5}$$

where Φ is the mode shape matrix and A represents the output displacement.

The analysis shows that the jth observational, x_j, data can be expressed as a linear combination of modal vectors ϕ_i; the a_{ij} are referred to as the coefficients associated with the ith eigenvector for the jth observational data and can be interpreted in the terms of temporal coefficients (Kutzbach 1967).

4.3.3 Physical Interpretation of the POD Modes

Proper orthogonal modes (POMs, EOFs) have an interesting analogy in terms of the modes of vibration of a mechanical system (Feeny and Kappagantu 1998). The key point to observe is that both, the POMs and the mechanical modes are orthogonal to each other; this suggests that in analogy with the mechanical counterpart, frequency (speed)-based signals might be used to approximate mode shapes (Messina and Vittal 2007).

To gain an insight into the physical interpretation of POMs, consider an n-dimensional mechanical system, $M\ddot{x} + Kx = 0$, where M and $K \in \mathfrak{R}^{n \times n}$ represent the mass (stiffness) matrix.

The free natural response is given by

$$x(t) = \sum_{j=1}^{n} \underbrace{A_j \sin(\omega_j + \theta_j)}_{a_j(t)} v_j \tag{4.6}$$

where the $v_j's$ are the natural modes, and the terms $a_j(t)$ represent the time modulation of the natural modes. Following the procedures outlined in Section 4.3.2, it can be shown that the eigenvectors (POMs), ϕ_j, of the correlation matrix, $R = X^T X$ converge to the modal vector v_j; the columns of the left eigenvector are the normalized time modulations of the eigen modes.

In the equations above, $a(t) \in \mathfrak{R}^{n \times 1}$ is the vector containing modal coordinates $A_j \sin(\omega_j + \theta_j)$ and $V \in \mathfrak{R}^{m \times m}$ is the mode shape matrix. Now, by direct analogy with (4.1), one can write

$$x(t) = Va(t) \tag{4.7}$$

where:

$$V = \begin{bmatrix} v_1 & v_2 \ldots & v_n \end{bmatrix}$$

$$a_j(t) = \begin{bmatrix} a_j(t_1) & a_j(t_2) \ldots & a_j(t_N) \end{bmatrix}$$

These techniques can be used to reduce the number of dimensions in measured data as well as to find patterns in the data. In addition, they also can be used to visualize data as discussed in the following sections.

4.3.4 SVD-Based Proper Orthogonal Decomposition

A useful alternative to conventional POD (EOF) analysis that avoids the computation of the covariance matrix, is obtained from singular value decomposition (SVD) analysis of the measurement matrix (Kerschen and Golinval 2002). Given a data matrix $X \in \Re^{m \times N}$ there exist two orthogonal matrices, $U \in \Re^{N \times N}$ and $V \in \Re^{m \times m}$, with $U^T U = I_N, V^T V = I_m$, such that

$$X = U \Sigma V^T = \begin{bmatrix} U & U_r \end{bmatrix} \begin{bmatrix} \Sigma & 0 \\ 0 & \Sigma_r \end{bmatrix} \begin{bmatrix} V_1^T \\ V_2^T \end{bmatrix} \approx \tilde{U} \tilde{\Sigma} \tilde{V}^T \tag{4.8}$$

where U and V are the left and right singular vectors of X, and $\Sigma \in \Re^{m \times M}$ is the matrix of singular vectors, with elements $\Sigma_{ii} = \sigma_{ii}$. From linear algebra, $\text{rank}(X) = r$, where r is the index of the small singular value.

If the low-rank approximation is exact, Σ_r is a matrix of zeroes and

$$\Sigma = \begin{bmatrix} \sigma_1 & & & \\ & \sigma_2 & & \\ & & \ddots & \\ & & & \sigma_r \end{bmatrix}$$

Letting $U_r = \begin{bmatrix} u_1 & u_2 \dots & u_r \end{bmatrix}$ and $V_r = \begin{bmatrix} v_1 & v_2 \dots & v_r \end{bmatrix}$ denote the singular vectors associated with the non-singular values, then one can write

$$X = \sigma_1 u_1 v_1^T + \sigma_2 u_2 v_2^T + \dots + \sigma_r u_r v_r^T = \sum_{j=1}^{r} \sigma_j u_j v_j^T \tag{4.9}$$

In terms of the mode shape matrix and modal coordinates, (4.9) can be rewritten in the form

$$X = UA(t) \tag{4.10}$$

where

$$A(t) = \begin{bmatrix} a_1(t) \\ a_2(t) \\ \vdots \\ a_r(t) \end{bmatrix} = \begin{bmatrix} a_1(t_1) & a_1(t_2)\dots & a_1(t_N) \\ a_2(t_1) & a_2(t_2)\dots & a_2(t_N) \\ \vdots & & \\ a_r(t_1) & a_r(t_2)\dots & a_r(t_N) \end{bmatrix}$$

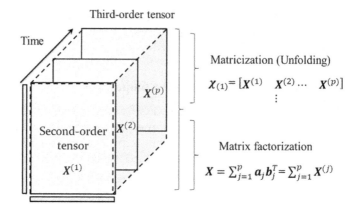

FIGURE 4.2
Tensor representation and unfolding (matricization) of data.

Comparison of (4.10) and (4.5) shows that $C = A$, and $\Phi = V$.

The *rhs* term in (4.9) has an interesting interpretation in terms of tensors, as shown in Figure 4.2. Using the properties $ab^T = a \otimes b$, the data matrix can be reconstructed as

$$X = a^{(1)} \otimes \psi_1 + a^{(1)} \otimes \psi_1 + \ldots + a^{(1)} \otimes \psi_1 = X^{(1)} + \ldots + X^{(p)} \tag{4.11}$$

where $X^{(k)} = a_k b_k^T$ and each term $\varphi_j \lambda_j a_j(t)$ represents a data matrix.

The equivalence between POD and other related techniques such as Principal Component Analysis (PCA) is discussed in Wu et al. (2003).

Two characteristics are of interest for the analysis of oscillatory problems:

1. The *m* spatial functions are mutually orthogonal

$$\sum_{i=1}^{m} \phi_i \phi_j = \delta_{ij} \tag{4.12}$$

2. The time dependent coefficients are mutually uncorrelated

$$\sum_{i=1}^{N} a_i(t) a_j(t) = \begin{cases} \lambda_i N, \, if \, i = j \\ 0, \, \text{otherwise} \end{cases} \tag{4.13}$$

The correlation matrix contains no information about the phase of the frequencies and represents the spectral content of different time scales and wavelengths (Sieber et al. 2016).

It can be shown that

$$
\begin{cases}
X = \overbrace{\Lambda_{av}}^{\substack{\text{Temporal} \\ \text{mean}}} + \overbrace{\Sigma_1 V_1^T}^{\substack{\text{Principal} \\ \text{components}}} \\[2em]
X = \underbrace{T_{av}}_{\substack{\text{Spatial} \\ \text{mean}}} + \overbrace{\Sigma_2 V_2^T}^{\substack{\text{Principal} \\ \text{components}}}
\end{cases}
\tag{4.14}
$$

where Λ_{av} and T_{av} are matrices containing the temporal and spatial patterns, and the products ΣV^T are called *principal components*.

By detrending the observation matrix, a first measure of power system oscillatory behavior is obtained from the time-demeaned matrix

$$
X_{osc} = X - \Lambda_{av}
\tag{4.15}
$$

Large deviations from the local mean are a first indicator of system deterioration. In addition, the shapes associated with critical modes provide information related to the extent and distribution of system damage. Mode shapes provide critical information for operational control actions (Trudnowski and Pierre 2009).

4.3.5 Principal Components Analysis

Principal Components Analysis (PCA) and its variants seek to create an orthogonal linear transformation that projects the original data into a new coordinate system in which the first dimension (or principal component) captures the largest possible amount of variance, the second dimension is orthogonal to the first while capturing the largest possible amount of remaining variance, and so on (Wold et al. 1987).

Formally, PCA transforms an mxN observation matrix, X, by combining the variables as a linear weighted sum

$$
X = \widehat{X} + TP^T + E = \widehat{X} + \sum_{i=1}^{a} t_i p_i^T + E
\tag{4.16}
$$

where $\widehat{X}$ is the center of X, $T \in \Re^{mxa}$ and $P \in \Re^{axN}$ are the scores and loading matrices, respectively; the matrix E contains error terms (Wold et al. 1987).

In terms of the loading and score matrices, one has that

$$
\begin{aligned}
X &= TP^T + E \\
Y &= UQ^T + F
\end{aligned}
\tag{4.17}
$$

and

$$U = TD$$

where T and U, denote orthonormal basis matrices of X-scores and Y-scores with dimension $n \times a$, and P and P are the X-loadings and Y-loadings, respectively; E and F are X-residuals and Y-residuals, respectively, and D is a diagonal matrix w_j.

As discussed in Chapter 5, the extent to which each principal component contributes to the particular waveform of ERP in the original data set is reported as a component score (Rosipal and Kramer 2005).

The main limitation of these approaches is their inability to explore associations and potential causal relationships between multiple data sets.

4.3.6 Koopman-Based Approximations

Another recently developed technique for spatiotemporal data mining and data-driven modeling, is Koopman Mode Decomposition (Susuki et al. 2016) and its variants: Dynamic Mode Decomposition (DMD) (Barocio et al. 2015; Hernández and Messina 2017) and the Extended DMD method.

Among these approximations, DMD creates a natural framework for data mining since the reduced dynamic model preserves the spatiotemporal dependence of the original feature space (Williams et al. 2015).

Given a model of the form (3.1), DMD allows the original model in physical space to be approximate by a model of the form

$$x_{j+1} = Ax_j, \qquad j = 0,\dots,N-1 \tag{4.18}$$

where A is a linear time-invariant operator (a time stepper) that maps the data from time t_j to time t_{j+1}.

For consistency with notation used in Sections 4.3.2–4.3.5, let the measurement matrix be defined as $X = \begin{bmatrix} x_1 & x_2 & \cdots & x_N \end{bmatrix} \in \Re^{m \times N}$. Dynamic mode decomposition estimates the Koopman operator. Using DMD analysis, singular value decomposition of the data matrix yields

$$x_1 = Ax_o, j = 0$$
$$x_2 = Ax_1 = A^2 x_o, j = 1$$
$$x_3 = Ax_2 = A^3 x_o, j = 2$$
$$\vdots$$
$$x_{N-1} = Ax_{N-2} = A^{N-1} x_o, j = N-2$$
$$x_N = Ax_{N-1} = A^N x_o, j = N-1$$

since Equation (4.18) should hold for all j.

Define now

$$X_o^{N-1} = \begin{bmatrix} x_o & x_1 & x_2 \cdots & x_{N-1} \end{bmatrix} \in \mathfrak{R}^{m \times N}$$

then,

$$X_1^N = A X_o^{N-1} = A \begin{bmatrix} x_0 & x_1 & x_2 \cdots & x_{N-1} \end{bmatrix} \tag{4.19}$$

or

$$X_1^N = \begin{bmatrix} x_1 & x_2 & x_3 \cdots & x_N \end{bmatrix} \in \mathfrak{R}^{m \times N}$$

It follows from (4.19) that the linear operator A can be obtained as

$$A = X_1^N \left(X_o^{N-1} \right)^+ \tag{4.20}$$

where $\left(X_o^{N-1} \right)^+$ denotes the pseudoinverse of X_o^{N-1}. Analysis of (4.20) is computationally demanding. In what follows, the above framework is extended to modeling large data sets.

This problem can be conveniently analyzed using singular value decomposition. From linear system theory, matrix X_o^{N-1} is diagonalizable and has a decomposition of the form

$$X_o^{N-1} = U\Sigma W^T = \begin{bmatrix} U_m & U_s \end{bmatrix} \begin{bmatrix} \Sigma_m & \\ & \Sigma_s \end{bmatrix} \begin{bmatrix} W_m^T \\ W_s^T \end{bmatrix} \tag{4.21}$$

with $U \in C^{m \times N}$, $W \in C^{m \times N}$, $UU^T = I$, $WW^T = I$, $r < N$, and (where $\Sigma_m \in \mathfrak{R}^{r \times r}$ includes the dominant singular values) Σ_s indicates the remaining $m-1-r$ singular values.

Equation (4.21) represents a low-rank approximation of X_1^N, and

$$X_o^{N-1} = W_m \Sigma_m^{-1} U_m^T \tag{4.22}$$

Substitution of (4.22) in (4.21) yields the approximation

$$\check{A} \approx A = X_1^N \underbrace{W_m \Sigma_m^{-1} \Sigma}_{\left(X_o^{N-1} \right)^+}$$

which can be rewritten as

$$X_1^N = A U_m \Sigma_m W_m^T$$

Multiplying this expression by $\mathbf{U}_m^T$ from the left and by $\mathbf{W}_m\Sigma_m^{-1}$ from the right yields

$$\mathbf{U}_m^T\mathbf{X}_1^N\mathbf{W}_m\Sigma_m^{-1} = \mathbf{U}_m^T\left(\mathbf{A}\mathbf{U}_m\Sigma_m\mathbf{W}_m^T\right) = \mathbf{U}_m^T\mathbf{A}\mathbf{U}_m = \tilde{\mathbf{S}} \in C^{m\times m}$$

Suppose now that matrix $\tilde{S}$ is diagonalizable with a decomposition of the form

$$\tilde{S} = Y\Lambda Y^{-1} \tag{4.23}$$

Substitution of (4.23) into (4.22) and solving for $\mathbf{X}_1^N$ yields

$$\mathbf{X}_1^N = \mathbf{U}_m Y\Lambda Y^{-1}\Sigma_m\mathbf{W}_m^T = \mathbf{U}_m\tilde{S}\Sigma_m\mathbf{W}_m^T$$

or

$$\mathbf{X}_1^N = \underbrace{\mathbf{U}_m}_{\substack{\text{Spatial}\\\text{structure}}} \underbrace{Y\Lambda Y^{-1}\Sigma_m\mathbf{W}_m^T}_{\substack{\text{Temporal}\\\text{structure}}} = \Phi\Lambda\Gamma_m(t) \tag{4.24}$$

where $\Phi = \mathbf{U}_m$,
   and

$$\mathbf{X}_1^N = \left[x_{dmd_1}(t) \quad x_{dmd_2}(t)\dots \quad x_{dmd_m}(t)\right] = \sum_{j=1}^{m}\phi_j\lambda_j a_j(t) \tag{4.25}$$

with

$$x_{dmd_j}(t) = \lambda_j\phi_j a_j(t)$$

where the quality of the approximation is given by $\left|\mathbf{X}_1^N - \widehat{\mathbf{U}}_m\tilde{S}\Sigma_m\mathbf{W}_m^T\right|$.

This relation is central to the development of spatiotemporal metrics of system behavior. One can show that (4.25) can be expressed in the alternate form

$$\mathbf{X}_{dmd} = \mathbf{X}_1^N = \sum_{i=1}^{r}b_i\varphi_i e^{(\sigma_i + j\omega_i)t} = \Phi e^{\Lambda t}b \tag{4.26}$$

where $b = \Phi^+x_0$.

It should be remarked that $\mathbf{X}_{dmd}^N$ has a similar interpretation to the model in Figure 4.3 and is amenable to tensor analysis. In analogy with Equation (4.9), each term, $\phi_j\lambda_j a_j(t)$ represents a matrix

$$\mathbf{X} = \mathbf{X}^{(1)} + \dots + \mathbf{X}^{(p)} = a^{(1)}\otimes\phi_1 + a^{(2)}\otimes\phi_2 + \dots + a^{(1)}\otimes\phi_p \tag{4.27}$$

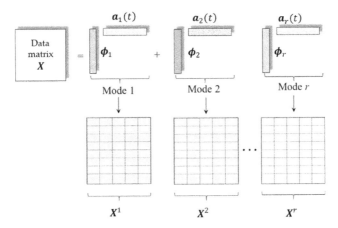

FIGURE 4.3
Tensor representation of the time series modal decomposition in (4.10).

where $\otimes$ is the Kronecker product, and $X^{(k)} = a^{(k)} \otimes \phi_k$.

A schematic representation of the model behind (4.27) is given in Figure 4.3. This model represents block matrix data structures that are amenable to sophisticated tensor analysis and parallel computing (Acar et al. 2019).

The DMD algorithm used in this presentation is adapted from Barocio et al. (2015) and Tu et al. (2014) and is briefly summarized in the boxed algorithm, Dynamic Mode Decomposition Algorithm:

Dynamic Mode Decomposition Algorithm

Given a data set $X = \begin{bmatrix} x_1 & x_2 & \cdots & x_N \end{bmatrix} \in \mathfrak{R}^{m \times N}$
1. Build matrices X_1^N and X_0^{N-1}.
2. Decompose the data matrix X using Equation (4.24).
3. Compute

$$\widehat{A} = U^* X^T W \Sigma^{-1}.$$

4. Compute the eigenvalues and eigenvectors of $\widehat{A}$ from

$$(\widehat{A} - \Lambda)W = 0.$$

5. Calculate the DMD modes as

$$\Psi = X^T V \Sigma^{-1} W.$$

6. Select the appropriate global scale $\varepsilon_i, \varepsilon_j$.
7. Calculate the spectral decomposition of $\widehat{P}$ and construct the low-dimensional mapping

$$\widehat{\Psi}(\widehat{X}) = \begin{bmatrix} \widehat{\Psi}(\widehat{X}^1) & \widehat{\Psi}(\widehat{X}^2) \cdots & \widehat{\Psi}(\widehat{X}^p) \end{bmatrix}.$$

8. Determine the temporal evolution of the Multiview model.

Remarks: Because the coordinates have a single frequency, the smaller scale fluctuations are represented by the higher order modes.

Two modifications to the standard DMD algorithm can enhance the ability of this method to mine large data sets. First, the reconstructed data matrix, X_{dmd}^N, is expressed as a sum of partial matrices associated with each term in (4.25)—Refer to Figure 4.3. Then the individual data blocks can be compared with similar representations from other data sets. Further, because DMD is a data reduction method, the method can be used to place dynamic recorders, based on observational data as discussed in Example 4.1.

Example 4.1

As an illustration of the use of spatiotemporal representations, a model of the form (4.16) and (4.27) is extracted from the time histories of speed deviations for the Australian test system. The basic purpose of this example is to find a simple representation that is amenable to tensor analysis.

According to (4.16) and assuming a 5-mode modal representation, the PCA decomposition of the observational data can be expressed in the form

$$\widehat{X} \approx X_1 + X_2 + X_3 + X_4 + X_5 \approx \sum_{j=1}^{5} \underbrace{t_j p_j^T}_{j} \tag{4.28}$$

using PCA (POD), or

$$\widehat{X} = X^{(1)} + X^{(2)} + X^{(3)} + X^{(4)} + X^{(5)} = \sum_{j=1}^{5} \underbrace{a^{(j)} \otimes \phi_j}_{x^{(j)}} \tag{4.29}$$

using DMD.

In (4.25), each component, $X^{(j)}, (X_j)$ represents a block matrix in Figure 4.3. Figure 4.4 depicts the generator speed computed using (4.25). Similar results are obtained using PCA and other time series analysis methods as discussed later in this book.

In spite of its simplicity, this example shows that models of the form (4.16) and (4.25) can be used to approximate system behavior using partial matrix (tensor) representations.

4.3.6.1 *Mode Selection*

Koopman mode decomposition and related techniques may result in a large number of modes. This makes modal interpretation difficult and may hinder physical interpretation.

In the case of DMD, the number of extracted modes is equal to the number of sensors. A question that naturally arises is how to determine whether the sensors or signals are numerous enough to accurately describe selected dynamics (i.e., slow system dynamics).

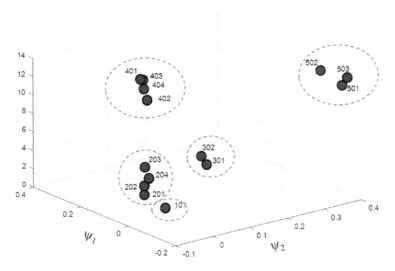

FIGURE 4.4
Reconstruction of speed signals of generator speeds computed using 3 modal components in (4.25).

TABLE 4.1

DMD Mode Selection Criteria

Approach	Criterion
Mode shape amplitude Barocio et al. (2015)	$\|\phi_i\| / \sum_j \|\phi_i\|$
Amplitude-based Alekseev et al. (2016)	$\|A_j\| / \sum_j \|A_j\|$
Energy-based Barocio et al. (2015)	$\|E_j\| / \sum_j \|E_j\|$

Several mode selection criteria—including amplitude, frequency or growth rate, and energy—have been proposed in the literature (Bistrian and Navon 2017; Barocio et al. 2015). Table 4.1. summarizes some approaches described in the literature to identify the most dominant modes of large dimensional trajectories. More sophisticated approaches are described in Ohmichi (2017) and Dang et al. (2018).

Experience shows that simple energy-based criteria are often used to select a subset of dominant modes. From physical considerations, the total fluctuating energy can be expressed as $E_j(t) = \|x_j(t)\|^2$. Noting that each mode is associated with a frequency ω_j and damping σ_j, Tissot et al. (2014) proposed the energy-weighted index

$$E_j = \frac{1}{T} \int_0^T \left\| \phi_j \lambda_j^{t/\Delta t} \right\|^2 dt = \frac{1}{T} \left\| \phi_j \right\|^2 \int_0^T \lambda_j^{2t/\Delta t} d \qquad (4.30)$$

where T is the time window of interest.

Noting that $e^{\log x} = x$ and integrating (4.30) with respect to T gives

$$E_j = \frac{1}{T}\left\|\phi_j\right\|^2 \int_0^T \lambda_j^{2\sigma_j t} dt = \left\|\phi_j\right\|^2 \frac{e^{2\sigma_j T} - 1}{2\sigma_j T} \tag{4.31}$$

It is apparent that modal energy is proportional to the square of the mode shape amplitude, the selected time interval T, and the damping decay σ_j. This makes this criterion useful to identify and isolate specific dynamic behavior associated with a given mode and time scale.

A second useful interpretation can be obtained by noting that the contribution of each state $x_i, i = 1,\ldots,n$ to mode j is given by

$$E_j(x) = \sum_{i=1}^{n} E_j(x_i) = \frac{e^{2\sigma_j T} - 1}{2\sigma_j T} \sum_{k=1}^{n}\left\|\phi_{ij}\right\|^2 \tag{4.32}$$

where ϕ_{ij} is the ith element of vector ϕ_j. This enables the development of efficient procedures to rank the importance of observation locations or system states. More general time dependent energy metrics are discussed in Section 4.3.6.2.

4.3.6.2 Spatiotemporal Clustering and Classification

A natural extension to DMD analysis is multivariate spatiotemporal clustering. Classification or discrimination between two or more groups of classes is also needed.

Once the dominant Koopman modes have been determined, clustering can be determined using the following boxed algorithm, Koopman-Based Clustering Strategy.

Koopman-Based Clustering Strategy

---

1. Determine the dominant modes and their associated eigenvectors $\Psi_k, k = 1,\ldots,r$ using information from equations (4.24) and (4.26).
2. Rank the modes in order of decreasing importance and discard singular vector showing a strong spatial correlation, using the criteria in equations (4.30) and (4.31).
3. Compute the clusters the dominant eigenvectors associated with dominant oscillatory modes (i.e., the most energetic modes exhibiting low-damping characteristics).

In order to implement this clustering technique, it is necessary to rank modal contributions in an efficient manner.

Example 4.2

To further illustrate these procedures, consider again the rotor speed deviations from example 3.2 in Chapter 3. The goal is to assess the applicability of various data mining techniques to extract coherent groups from time series of speed deviations using various clustering data mining techniques.

Three approaches to clustering rotor speed deviations are considered:

Procedure 1. *C*-means clustering was applied directly to measured rotor speed deviations.

Procedure 2. *K*-means clustering was applied to the low-dimensional embedding obtained using diffusion maps. In this case, the eigenvectors of the low-dimensional representation are written the form

$$X = \begin{bmatrix} \phi_1 \, \phi_2 \ldots \phi_d \end{bmatrix} = \begin{bmatrix} \phi_{11}^{(1)} & \phi_{11}^{(2)} & \phi_{11}^{(d)} \\ \phi_{21}^{(2)} & \phi_{21}^{(2)} & \phi_{21}^{(d)} \\ \vdots & \vdots & \vdots \\ \phi_{m1}^{(1)} & \phi_{m1}^{(2)} & \phi_{m1}^{(d)} \end{bmatrix}$$

Procedure 3. Clusters are obtained by projecting the two dominant eigenvectors extracted using DMD analysis.

Application of DMD to the rotor speed deviations in Figure 4.5, with $k = 7$ results in 14 dynamic modes, while the application of Koopman mode analysis results in 1703 modes.

Table 4.2 shows the damping and frequency of the extracted modes, along with the associated modal energy. The system response exhibits two dominant modes ($d = 2$, in procedure 2) at 1.39 Hz (a local mode) and a 0.381 Hz (inter-area mode 2 in Table 3.5). In these results, the modal damping and frequency are computed from

$$\rho_j = \Re\left\{\left[\log\left(\lambda_j\right)\right]\right\} / \Delta t$$
$$f_j = Im\left\{\left[\log\left(\lambda_j\right)\right]\right\} / \Delta t / 2\pi \tag{4.33}$$

Table 4.3 lists the dominant modes extracted using KMD, while Figure 4.4 shows the extracted coherent groups using this approach. As shown in this plot, projection of the two dominant eigenvectors associated with modes 1 and 2 in Table 4.1 identifies five significant spatiotemporal clusters in close agreement with observed system behavior.

4.3.6.3 Application as a Predictive Tool

Recently, DMD has been used as a predictive tool (Lu and Tartakovsky 2019). Having computed the eigenvalues and eigenvectors of matrix $\hat{A}$ in the DMD algorithm, a prediction into future behavior can be obtained as

$$X_{dmd}^{j+1} = \Phi\Lambda^{j+1}b, j > N \tag{4.34}$$

for $j > N$. The simplicity of this approach makes it specially well suited for short-term prediction.

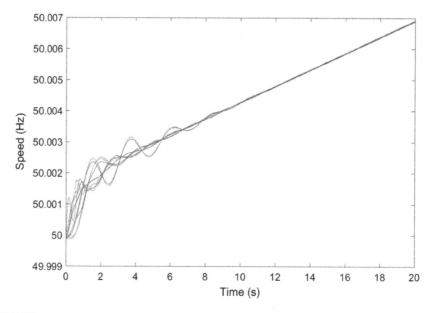

FIGURE 4.5

Second eigenvectors versus first eigenvector for clustering of rotor speed deviations in Figure 3.11. Procedure 3.

TABLE 4.2

Koopman mode analysis for the rotor speed deviations following a 1% load change at bus 217. Australian test power system.

Mode	Eigenvalue	Frequency (Hz)	Damping (%)	Modal Energy
1	$-0.5278 \pm j1.7965$	1.391	2.77	0.654
2	$-0.3809 \pm j2.3994$	0.391	27.71	0.313

TABLE 4.3

K-Means Clustering of Rotor Speed Deviations. Procedures 1 and 2

Cluster	Procedure 1	Procedure 2
1	402, 404, 403	404, 402, 403
2	401	401
3	501, 503, 502	501, 503, 502
4	201, 202, 204, 203	201, 202, 204, 203
5	301, 302	301, 302

4.4 Space-Time Prediction and Forecasting

Modern wide-area measurements and control systems rely on space-time prediction and forecasting systems to detect impending threats and start early warning systems. Estimation of global behavior using a limited number of measurements is a major research topic across various fields of study (Messina 2009 and Zaron 2011).

4.4.1 Interpolation and Extrapolation

A major challenge in space-time prediction analysis is to estimate the power system state using the available information at observed locations. Depending on the nature of the array of sensors and system coverage two approaches are possible: spatial interpolation and spatial extrapolation.

Spatial interpolation is defined as the prediction of the unknown value of a physical variable from known measurements obtained at a set of sample locations (Munoz et al. 2008). The application of these ideas to measured power system data is discussed in Messina (2015) and is examined in Chapter 6.

Much as in the case of other observing systems in other fields (Zaron 2011; Messina 2015), three main cases can be considered:

1. Dense or redundant measurements at a local level,
2. Sparse measurements at a local or global level, and
3. Lack of measurements.

In the first case, measurements are expected to be highly correlated. Typically, some sort of model reduction or feature extraction/selection is needed to avoid redundant information as well as to extract the underlying dimensions of the data. Smoothing or data filling techniques are also needed in the case of missing data or dropped data points (Allen et al. 2013).

In the case of sparse sensor networks or measurements, spatiotemporal interpolation (prediction) may be used to estimate system behavior at unsampled (unmonitored) system locations and fill in missing data. *Extrapolation* may also be needed to estimate values of physical variables in remote locations where they are not measured.

Proper analysis of these models requires that space and time are combined or integrated in a proper manner (Cheng et al. 2013).

In the analysis that follows, the wide-area monitoring system is developed as a data assimilation system and possesses the following capabilities:

- Selecting a subset of measurements or variables from the original data set (variable or feature selection);
- Extracting from the observed response, the key dynamics of interest;

- Estimating system behavior at unmonitored system locations (unsampled points in space) and smoothing system measurements; and
- Predicting system behavior at future time.

The first problem can be addressed using a suitable dimensionality reduction/feature extraction technique and optimal sensor location techniques (Castillo and Messina 2019).

4.4.2 Compressive Sampling

Let x be an N—dimensional signal that is to be measured. Assume, further, that x has a sparse representation in an orthonormal basis $\Psi = \begin{bmatrix} \psi_1 \, \psi_2 \ldots \psi_N \end{bmatrix}$. It follows from the theory in Section 4.3, that a signal, x, can be expressed in terms of linear basis as

$$x = \sum_{i=1}^{m} s_i \psi_i \tag{4.35}$$

or $x = \Psi s$,

where $s = \begin{bmatrix} s_1 \, s_2 \ldots s_N \end{bmatrix}^T$ is a vector of weighting coefficients. Noting that $\Psi^T \Psi = I$, it immediately follows that $s = \Psi^T x$, and $s_i = \psi_i^T x$.

Intuitively, the signal is K-sparse if it can be represented as a linear combination of only K vectors of the orthonormal basis, with $K \ll N$. This means that only K of the s_i coefficients are nonzero. In other words, the signal x is compressible if the representation (4.29) has just a few large coefficients and many small coefficients (Ganguli and Sompolinsky 2012, Bao et al 2019, Donoho 2006).

In order to illustrate the use of these techniques, consider again the measurement matrix, X

$$X = \begin{bmatrix} x_1(t_1) & x_1(t_2) & \cdots & x_1(t_N) \\ x_2(t_1) & x_2(t_2) & \cdots & x_2(t_N) \\ \vdots & \vdots & \ddots & \vdots \\ x_m(t_1) & x_m(t_2) & \cdots & x_m(t_N) \end{bmatrix}$$

Now consider a general measurement vector (a reduced order representation) $y \in \Re^M$, with $M < N$ and assume that $y = Cx$, where $C \in \Re^{M \times N}$ is a measurement matrix. Substituting (4.35) into this expression yields

$$y = Cx = C\psi s = \Theta s \tag{4.36}$$

where $\Theta \in \Re^{M \times N} = C\Psi$.

Two problems are of interest here:

1. Finding a measurement matrix, **C** such that the relevant information in any K sparse signal is not modified by the reduction process, and
2. Determining a reconstruction algorithm to recover x from $M \approx K$ measurements y (Baraniuk 2007).

An illustration of the problem of compressive sampling is given in Figure 4.6. This problem can be cast into the problem of a compressive sensing-based machine learning problem.

A closely related problem is that of determining a binary measurement matrix that allows to reconstruct a measured signal using a few dominant modes (Zhang and Bellingham 2008). The discussion of this issue is postponed to Chapter 6.

Prediction requires the integration of temporal attributes such as the extraction of time-varying means and oscillatory components as well as some physical or spatial information.

Spatiotemporal prediction and forecasting techniques include methods such as artificial neural networks, support vector machines (SVMs), and support vector regression (SVR), among others (Farantatos and Amidan 2019).

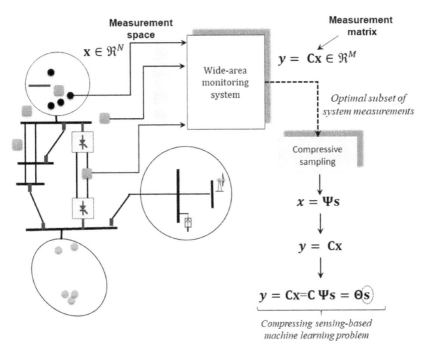

FIGURE 4.6
Schematic illustration of dimensionality reduction. (Based on Baraniuk 2007.)

In several practical applications, spatial prediction is usually obtained using neighboring stations in a network of a dense grid. In the more general case, spatiotemporal prediction of a dynamic process at a non-sampled station at time can be obtained from the time evolution of selected measurements.

4.4.3　Spatiotemporal Clustering and Classification

One of the main applications of POD and PCA has been to identify a low-dimensional space that best explains the data's spatiotemporal variance. By taking the data's first principal components, it is possible to identify dominant spatial structures and their evolution over time. Clustering techniques can also be used as a data mining tool to analyze and compare very large data sets. In some cases, it is possible to link the dominant POD modes to electromechanical modes (Messina and Vittal 2007). Similar approaches have been employed other techniques such as ICA and Blind Source Separation (Nuño Ayón et al. 2015), to mention a few techniques.

4.5　Spatiotemporal Data Mining and Pattern Evaluation

4.5.1　Pattern Mining

Models that describe the relation between two variables have been recently used to mine power system data. Examples include Partial Least Squares Correlation (PLSC), Canonical Correlation Analysis (CCA), and the hybrid models Canonical Correlation Partial Least Squares (CCPLS) (Kruger and Qin 2003). These techniques can be used to identify dynamic patterns of system behavior and often provide complementary information. They are especially useful in mining spatiotemporal data to identify combinations of variables that are maximally correlated.

Partial least square regression (PLS) has been recently used for pattern mining of measured data (Messina 2009; Wold et al. 2010). Two useful extensions to PLS are Partial Least Squares Correlation (PLSC) and Partial Least Squares Regression (PLSR). In the first case, the goal is to find the patterns of association between two data sets; in the second, interest is focused on predicting one data set from the other. This information is of interest to determine optimal relationships between two sets of data; for example, voltage and reactive power or frequency and active power. Other approaches seeking to analyze relationships between variables are CCA and the hybrid method canonical PLS (Kruger and Qin 2003).

These two methods are briefly outlined in Sections 4.5.2 and 4.5.3 as an extension to other more general formulations.

4.5.2 Spatiotemporal Autocorrelation

Partial Least Squares Correlation (PLSC) analyzes the relationship between two matrices X and Y. Following (Van Roon et al. 2014; Abdi and Williams 2013), the correlation between data sets X and Y is defined as

$$R = Y^T X \qquad (4.37)$$

where R is called the cross-block covariance matrix, and both X and Y are assumed to be centered and normalized.

Singular value decomposition analysis of (4.14) yields

$$R = Y^T X = U \Lambda V^T \qquad (4.38)$$

where U and V are the matrices of normalized eigenvectors of RR^T and are called *saliences*, and Λ is the diagonal matrix with square root of eigenvalues.

The latent variables L_X and L_Y are obtained by projecting the data sets X and Y yields onto their saliencies as

$$L_x = XV$$
$$L_y = YU$$

Physically, matrices U and V can be used to determine swing patters in measured data, while the correlation matrix R provides a measure of interaction between data sets X and Y. The extension of this approach to the multivariate case allows simultaneous analysis of cross-information between multiple data sets.

Drawing on this model, multivariate statistical tools to systematically analyze multiblock data structures across a wide range of spatial and temporal scales using multiblock PLSC and regression can be defined as discussed in Section 4.5.3.

4.5.3 Canonical Correlation Analysis

A canonical correlation analysis is a generic parametric model used in the statistical analysis of data involving interrelated or interdependent input and output variables. One shortcoming of the canonical correlation analysis, however, is that it provides only a linear combination of variables that maximizes these correlations.

Given two data sets $X \in \mathfrak{R}^{N \times m}$ and $Y \in \mathfrak{R}^{N \times n}$, CCA seeks a pair of linear transforms $w_x \in \mathfrak{R}^{N \times 1}$ and $w_y \in \mathfrak{R}^{m \times 1}$, such that the correlation between linear combinations $\bar{x} = w_x^T X$ and $\bar{y} = w_y^T X$ is maximized (Kruger and Qin 2003; Yair and Talmon 2017)—namely

$$\max_{\underbrace{\rho}_{w_x,w_y}} = \frac{w_x^T XY^T w_y}{\sqrt{w_x^T XX^T w_x w_y^T YY^T w_y}} \tag{4.39}$$

where the maximum of the correlation coefficient ρ, with respect to w_x and w_y is called the *maximum canonical correlation*.

The following example examines the application of various spatiotemporal analysis techniques to characterize and extract voltage and reactive power patterns from observational data.

Example 4.3

Figure 4.7 shows a portion of a large system used to investigate the application of dimensionality reduction techniques to monitor and characterize voltage and reactive power patterns. The observation set contains transient stability data and covers a period of 20 seconds observed at 174 locations. The set includes bus voltage magnitudes and reactive power data. The primary objective is to fuse the data together to produce a single data set for analysis and monitoring of reactive control areas.

Five data sets are used to illustrate the complementary nature of multimodal data POD to identify dominant structures. Contingency scenarios include load-shedding at major load buses and three-phase faults. The observational data consists of 176 bus voltage magnitudes from selected transmission buses in a regional system. In each case, the

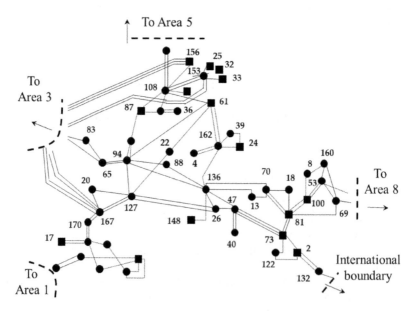

FIGURE 4.7
Schematic of the study system.

feature vector is defined as $X_k = \begin{bmatrix} V_1 & V_2 \ldots & V_{174} \end{bmatrix}^T, k = 1, \ldots, 5$, where $V_j(t) = \begin{bmatrix} V_j(t & V_2 \ldots & V_t \end{bmatrix}^T, k = 1, \ldots, r$, with $r = 5$.

Direct concatenation of the data sets may be problematic because of ill-conditioning of the methods. A simple technique is to average the data sets. This approach is reasonable when data mining or data fusion techniques are applied to very similar data sets.

After this pre-processing, nonlinear dimensionality reduction techniques were applied to the fused data sets. Dynamic mode decomposition results in 174 dynamic modes.

The number of clusters, m, was chosen based on previous application of projection-based clustering techniques. The results show 10 well defined clusters with a strong physical correlation (Table 4.4).

Having determined the cluster partitions and centroids (non-physical nodes), an approximation to the notion of pilot centers can be determined by calculating the minimum distance between the buses in each area and its centroid. For each candidate set L of clusters, the intercluster distance $D(c_i)$, between the centroids and the cluster members $x \in c_i$, can be calculated as

$$D(c_i) = \frac{1}{n} \sum_{x \in c_i} (x - C_i) \tag{4.40}$$

Refer to Section 3.5 for details.

This demonstrates that the combination of multiple feature selection methods can improve the robustness and accuracy of detection.

An alternative approach consists of applying spectral methods to the raw data. In this case, application of DMs to m time series requires the

TABLE 4.4

Clustering of Bus Voltage Magnitudes Using K-Means and c-Means. $m = 10$.

Cluster	K-Means	C-Means
1	1–7, 10–12, 15–17, 20–24, 26–28, 30–31, 34, 38–52, 57–65, 67–68, 73–77, 83–86, 88–94, 97–99, 101–104, 116–127, 125–131, 135–136, 139–152, 162–173	1, 4–7, 10–11, 17, 20, 24, 39, 60–64, 73, 75–77, 83–84, 127–128, 130, 151–152, 162–173
2	107–115	107–115, 137–138
3	8–9, 53–56, 66, 69, 100, 160–161	8–9, 53–56, 66, 69, 78–80, 82, 100, 160–161
4	105–106	---
5	29, 87, 96, 156–159	36, 87, 96, 156–159
6	25, 32–33, 36–37, 153–155	25, 32–33, 153–155
7	2–3, 35, 124, 132–134	2–3, 124, 132–134–134
8	95, 137–138	95, 105–106
9	13–14, 18–19, 70–72, 81	13–14, 18–19, 70–72, 74, 81
10	78–80, 82	12, 29–31, 35, 37, 57–59, 85–86, 101–104, 117–123

"---" included in cluster 8.

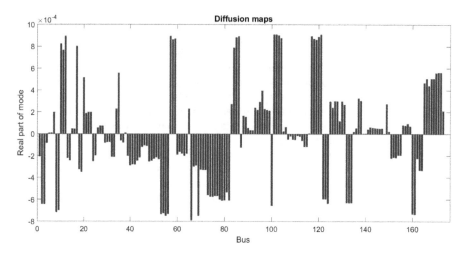

FIGURE 4.8
Voltage-based mode shape.

calculation $m(m-1)/2$ pairwise distances. This makes the application of the method computationally demanding (especially as the number of sensors increases), but provides additional information about modal behavior.

Figure 4.8 shows the voltage-based shape Ψ_V extracted using diffusion maps. Using this information, modal behavior can be obtained from eigenanalysis of the state matrix.

References

Abdi, H., Williams, L. J., Partial least squares methods: Partial least squares correlation and partial least square regression, Chapter 23, in Reisfeld B., and Mayeno A. N. (Eds.), *Computational Toxicology: Volume II, Methods in Molecular Biology*, Vol. 930, pp. 549–578, Humana Press, Totowa, NJ, 2013.

Acar, E., Kolda, T. G., Dunlavy, D. M., All-at-once optimization for coupled matrix and tensor factorizations, ArXiv.org, 1105.3422, e-print, May 2011.

Alekseev, A. K., Bistrian, D. A., Bondarev, A. E., Navon, I. M., On linear and nonlinear aspects of dynamic mode decomposition, *International Journal for Numerical Methods in Fluids*, 82, 348–371, 2016.

Allen, A., Santoso, S., Muljadi, E., Algorithm for screening phasor measurement unit data for power system events and categories and common characteristics for events seen in phasor measurement unit relative phase-angle differences and frequency signals, Technical Report, NREL/TP-5500-58611, Golden, CO, August, 2013.

Bao, Y., Tang, Z., Li, Z., Compressive-sensing data reconstruction for structural health monitoring: A machine learning approach, arXiv:1901.01995, March 2019.

Baraniuk, R. G., Lecture notes, *IEEE Signal Processing Magazine*, 118–124, July 2007.

Barocio, E., Pal, B. C., Thornhill, N. F., Roman-Messina, A., A dynamic mode decomposition framework for global power system oscillation analysis, *IEEE Transactions on Power Systems*, 30(6), 2902–2912, 2015.

Bistrian, D. A., Navon, I. M., The method of dynamic decomposition in shallow water and a swirling flow problem, *International Journal for Numerical Methods in Fluids*, 83(1), 73–89, 2017.

Castillo, A., Messina, A. R., Data-driven sensor placement for state reconstruction via POD analysis, *IET Generation, Transmission & Distribution*, 14(4), 656–664, 2020.

Cheng, T., Haworth J., Anbaroglu, B., Tanaksaranond, G., Wang, J., *Spatio-Temporal Data Mining*, Book Chapter, Springer Verlag, Berlin, Germany, 2013.

Chu, W. W. (Eds.), *Data Mining and Knowledge Discovery for Big Data – Methodologies, Challenges and Opportunities*, Springer-Verlag, Berlin, Germany, 2014.

Dang, Z., Lv, Y., Li, Y., Wei, G., Improved dynamic mode decomposition and its application to fault diagnosis of rolling bearing, *Sensors*, 18, 1972, 1–15, 2018.

Donoho, D. L., Compressed sensing, *IEEE Transactions on Information Theory*, 52(4), 1289–1306, 2006.

Esling, P., Agon, C., Time-series data mining, *ACM Computing Surveys*, 45(1), 12:1–12:34, 2012.

Farantatos, E., Amidan B. (Eds.), *NASPI White Paper: Data Mining Techniques and Tools for Synchrophasor Data, Prepared by NASPI Engineering Analysis Task Team (EATT)*, PNNL-28218, January 2019.

Fayyad, U., Piatetsky-Shapiro, G., Smyth, P., From data mining to knowledge discovery in databases, *AI Magazine*, 17(3), 37–53, 1996.

Feeny, B. F., Kappagantu, R., On the physical interpretation of proper orthogonal modes in vibrations, *Journal of Sound and Vibration*, 211(4), 607–616, 1998.

Ganguli, S., Sompolinsky, H., Compressed sensing, sparsity, and dimensionality in neuronal information processing and data analysis, *Annual Review in Neuroscience*, 35, 485–508, 2012.

Ghamisi, P., Rasti, B., Yohoka, N., Wang, Q., Hofle, B., Bruzzone, L., Bovolo, F. et al., Multisource and multitemporal data fusion in remote sensing: A comprehensive review of the state of the art, *IEEE Geoscience and Remote Sensing Magazine*, pp. 6–39, March 2019.

Hand, D., Mannila, H., Smyth, P., *Principles of Data Mining*, The MIT Press, Cambridge, MA, 2001.

Hannachi, A., Jolliffe, I.T., Stephenson, D. B., Empirical orthogonal functions and related techniques in atmospheric science: A review, *International Journal of Climatology*, 27, 1119–1152, 2007.

Harvey, A. C., *Forecasting, Structural Time Series Models and the Kalman Filter*, Cambridge University Press, London, UK, 1990.

Hernández, M. A., Messina, A. R., An observability-based approach to extract spatiotemporal patterns from power system Koopman mode analysis, *Electric Power Components and Systems*, 45(4), 355–365, 2017.

Kerschen, G., Golinval, J. C., Physical interpretation of the proper orthogonal modes using the singular value decomposition, *Journal of Sound and Vibration*, 249(5), 849–865, 2002.

Kruger, U., Qin, S. J., Canonical correlation partial least squares, *IFAC Proceedings*, 36(16), 1603–1608, 2003.

Kutzbach, J. E., Empirical eigenvectors of sea-level pressure, surface temperature and precipitation complexes over north America, *Journal of Applied Meteorology*, 6, 791–802, 1967.

Liu, G., Quintero, J., Venkatasubramanian, V., Oscillation monitoring system based on wide area synchrophasors in Power Systems, *2007 IREP Symposium, Bulk Power Systems Dynamic and Control*, Charleston, South Carolina, August 2007.

Lu, H, Tartakovsky, T., Predictive accuracy of dynamic mode decomposition, arXiv preprint arXiv:1905.01587, 2019.

Messina, A. R. (Eds.), *Inter-Area Oscillations in Power Systems, A Nonlinear and Nonstationary Perspective*, Springer, New York, 2009.

Messina, A. R., Vittal, V., Extraction of dynamic patterns from wide-area measurements using empirical orthogonal functions, *IEEE Transaction on Power Systems*, 22(2), 682–692, 2007.

Munoz, M., Lesser, V. M., Ramsey, F. L., Design-based empirical orthogonal function model for environmental monitoring data analysis, *Environmetrics*, 19, 805–817, 2008.

Nuño Ayón, J. J., Barocio, E., Roman-Messina, A., Blind extraction and characterization of power system oscillatory modes, *Electric Power Systems Research*, 119, 54–65, 2015.

Ohmichi, Y., Preconditioned dynamic mode decomposition and mode selection algorithms for large datasets using incremental proper orthogonal decomposition, arXiv:1704.03181, April 2017.

Rosipal, R., Kramer, N., Overview and recent advances in partial least squares, 34–51, February, 2005, in Saunders, C., Grobelnik, M., Gunn, S., Shawe-Taylor, J. (Eds.), *Subspace, Latent Structure and Feature Selection, Lecture Notes in Computer Science*, Springer, Berlin, Germany, February, 2005.

Sarmani, S. A. N., Venkatasubramanian, V., Inter-area resonance in power systems from forced oscillations, *IEEE Transaction on Power Systems*, 31(1), 378–386, 2016.

Shekhar, S., Jiang, Z., Ali, R. Y., Eftelioglu, E., Tang, X., Gunturi, V. M. V., Zhou, X., Spatiotemporal data mining: A computational perspective, *ISPRS International Journal of Geo-Information*, 4, 2306–2338, 2015.

Sieber, M., Paschereit, C. O., Oberleithner, K., Spectral proper orthogonal decomposition, *Journal of Fluid Mechanics*, 792, 798–828, 2016.

Susuki, Y., Mezic, I., Raak, Hikihara, T., Applied Koopman operator theory for power systems technology, *Nonlinear Theory and Its Applications*, 7(4), 430–459, 2016.

Tissot, G., Cordier, L., Benard, N., Noack, B. R., Model reduction using dynamic mode decomposition, *Comptes Rendus Mécanique*, 342(6–7), 410–416, 2014.

Trudnowski, D., Pierre, J., Signal processing methods for estimating small-signal dynamic properties from measured responses, in Messina, A. R. (Ed.), *Inter-Area Oscillations in Power Systems*, Springer Verlag, New York, 2009.

Tu, J., Rowley, C., Luchtenberg, D., Brunton, S., Kutz, J. N., On dynamic mode decomposition: Theory and applications, *Journal of Computational Dynamics*, 1, 391–421, 2014.

Van Roon, P., Zakizadeh, J., Chartier, S., Partial least squares tutorial for analyzing neuroimaging data, *The Quantitative Methods for Psychology*, 10(2), 200–215, 2014.

Webber, G. A., Handler, R. A., Sirovich, L., Energy dynamics in a turbulent channel flow using the Karhunen-Loeve approach, *International Journal for Numerical Methods in Fluids*, 40, 1381–1400, 2002.

Williams, M. O., Rowley, C. W., Mezic, I., Kevrekidis, I. G., Data fusion via intrinsic dynamic variables: An application of data-driven Koopman spectral analysis, *EPL (European Physics Letters)*, 109(4), 2015.

Wold, S., Eriksson, L., Kettaneh, N., PLS in data mining and data integration, Chapter 15, in Vinzi, V. E., Chin, W. W., Henseler, J., Wang, H. (Eds.), *Handbook of Partial Least Squares, Concepts, Methods and Applications*, Springer Verlag, Berlin, Germany, 2010.

Wold, S., Esbensen, K., Geladi, P., Principal component analysis, *Chemometrics and Intelligent Laboratory Systems*, 2, 37–52, 1987.

Wu, C. G., Liang, Y. C., Lin, W. Z., Lee, H. P., Lim, S. P., A note on equivalence of proper orthogonal decomposition methods, *Journal of Sound and Vibration*, 265(5), 1103–1110, 2003.

Yair, O., Talmon, R., Local canonical correlation analysis for nonlinear common variables discovery, *IEEE Transactions on Signal Processing*, 65(5), 1101–1115, 2017.

Zaron, E. D., Introduction to Ocean data assimilation, Chapter 13, in Schiller, A., Brassington, G. B. (Eds.), *Operational Oceanography in the 21st Century*, Springer, Dordrecht, the Netherlands, 2011.

Zhang, Y., Bellingham, J. G., An efficient method for selecting ocean observing locations for capturing the leading modes and reconstructing the full field, *Journal of Geophysical Research*, 113, C04005, 1–24, 2008.

5

Multisensor Data Fusion

5.1 Introduction and Motivation

Joint data analysis and fusion of multivariate data is key to modern forecasting and prognosis systems. Fusion of data from multiple sources (multisensor or multiview data fusion) can help identify interactions across various measurement types or be used to provide complementary information in an integrated manner (Hall and Llinas 1997; Dalla Mura et al. 2015; Allerton and Jia 2005).

Data fusion converts raw data into useful information, which is subsequently combined with knowledge and logic to aid in inference and decision making. To achieve this, data fusion may incorporate other techniques such as data mining, data correlation, and assimilation techniques.

Measured, power system spatiotemporal data presents unique challenges to data fusion techniques because measured records can differ in many ways (Kezunovic and Abur 2005; Zhang et al. 2012; Ghamisi et al. 2019). Further, power system dynamic behavior is becoming more variable and uncertain due to changes in generation infrastructure and a less predictable load and generation behavior. Higher penetration of distributed resources along with the increased utilization of diverse recording devices is making data fusion more challenging (Fusco et al. 2017).

Addressing these issues requires the development of new tools to systematically accommodate data collected across a wide range of spatial and temporal scales (Abdulhafiz and Khamis 2013). Advanced fusion techniques should be well suited to automation, be robust, and incorporate both prediction and prognosis features to anticipate harmful conditions.

Advantages of multisensor data fusion include improved situation assessment and awareness, improved detection and tracking, improved robustness, and extended spatial and temporal coverage. Another area benefitting from application of data fusion is anomaly detection. These are issues that have been scarcely investigated in the analysis of measured power system time series.

In the context of power system monitoring, data fusion methods are adopted to combine complementary information from measurements and models to determine a more coherent and accurate representation of system behavior. This information can then be used for prognosis and decision (Schmitt and Zhu 2016). In addition, data fusion techniques have the potential to make the system safer by increasing the accuracy and reliability of wide-area measurements.

Several frameworks for fusing data from multiple sensors have been proposed in the literature. Typical fusion techniques for multisensor inferencing include signal detection and estimation theory, Kalman filtering, estimation and filtering, neural networks, clustering, fuzzy logic, knowledge-based systems, and control and optimization algorithms to name a few. Other recent approaches include approximations to Koopman eigenfunctions (Williams et al. 2015).

In the study of multisource and multivariate data, a framework is needed for fusing data and extracting not only reduced-order models, but also possible cross-relationships among items in the series. In this chapter, a review of multisource data fusion methods from the perspective of power system monitoring is provided. Multivariate statistical techniques based on Partial Least Squares (PLS) designed for determining and modeling dynamic relationships between multivariate data blocks are also introduced. These have been tested on synchrophasor data to identify changing patterns between data, detect disturbances, and track mode propagation on a global scale.

5.2 Spatiotemporal Data Fusion

Many problems in power system data analysis involve the analysis of multiple data sets, either of the same type (i.e., frequency deviations) as in the case of the power system response to different perturbations, or where the analysis involves measurement of different data types at different system locations. Such multivariate multitype data arise when dynamic phenomena are observed simultaneously using WAMS or other sensor networks.

Data-driven fusion of multimodality data is an especially challenging problem since power system dynamic data types are intrinsically dissimilar in nature. In addition, multimodal data may have different features (such as trends) or be sensitive to outliers or noise. In many applications involving wide-area monitoring and control, it is desirable (and often advantageous) to monitor and fuse data from different modalities. For example, it may be useful to know the spatial relationship between measured data (Dutta and Overbye 2014). To accomplish this, data need to be combined (fused) to create a single spatiotemporal measurement data set in order to obtain complementary information.

In what follows, methods for extracting the dominant temporal and spatial components of variability in measured data are investigated and numerical issues are addressed. Recent progress on the development of multisensory data fusion techniques are highlighted and reviewed in the context of power system monitoring.

5.3 Data Fusion Principles

Data fusion is defined by some authors as the process of integrating data from a number of sensors or sources to make a more robust and confident decision about a process of interest than is possible with any single sensor alone (Worden et al. 2011). Extracting joint information from multiple data sets can be accomplished using univariate approaches (such as correlation) or multivariate approaches as discussed in Sections 5.3.2 and 5.3.3.

5.3.1 Problem Formulation

Exploratory methods that simultaneously model both shared features and features that are specific to each data source have recently been developed. To introduce the notion of multisensor data fusion, consider a set of spatiotemporal matrices $X^l \in \mathfrak{R}^{N \times m_l}$, $l = 1, \dots, L$, with $L \geq 2$, where N denotes the number of data samples and m_l represents the number of sensors associated with each particular data set. These data sets could represent homogeneous (single-type) data measured at different control areas, or inhomogeneous data collected on the same set of sensors (i.e., voltage, reactive power, voltage phase magnitudes). The latter problem is of interest here.

The goal is to fuse the heterogeneous data matrices $X^1, X^2, \dots, X^L$ by computing a low-dimensional representation $\hat{X}$. Many methods to compare multiple large-scale matrices are limited to two matrices. Thus, for instance, for the two-dimensional (two-block) case, one can concatenate the data matrix into the multiblock data matrix $X = [X^1 \; X^2]$. Next, extensions to the multidimensional case are explored.

Drawing on these ideas, the heterogeneous data sets, X^l, $l = 1, \dots, L$, can be directly combined and concatenated by rows (or columns) into a single data matrix

$$\hat{X} = \begin{bmatrix} X^1 & X^2 \dots & X^L \end{bmatrix} \in \mathfrak{R}^{N \times m_T} \tag{5.1}$$

where $m_T = m_1 + \dots + m_L$. Other representations and interpretations concerning merging of heterogeneous data are possible and are described in references (Messina 2015; Savopol and Armenakis 2002). A schematic of this representation is shown in Figure 5.1; the left-hand side of the plot shows a representation of the data sets, while the right panel gives a representation of the concatenated form (5.1). This process is referred to as *unfolding*.

However, it is well known that direct analysis of models of the form (5.1) may be difficult as the dimensionality and nature of the data sets may be different or they may have different properties or units, which requires proper scaling (Lock et al. 2013). Moreover, analyses of block-by-block correlations

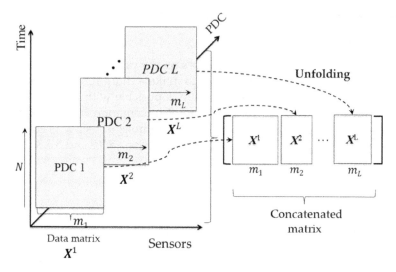

FIGURE 5.1
Concatenated data matrix $\widehat{X}$. For clarity of illustration, data sets $X^1, ..., X^L$ are assumed to have the same number of rows (sensors) but different number of samples (columns). (Adapted from Krishnan, A., et al., *NeuroImage*, 56, 455–475, 2011.)

cannot identify the global modes of variation that drive associations across and within data types. This means that interactions between data sets of different modalities cannot be accounted for by directed analysis of the extended, unfolded matrix.

In general, two basic approaches to data fusion are found in the literature:

1. Data fusion of a concatenated data matrix of the form (5.1).
2. Separate analyses of the various data sets followed by superposition of individual fusion results.

In the cases that follow, a framework for data fusion is introduced. Four basic cases of multisensor data fusion are of interest here and are considered in these derivations:

Case 1. Fusion of multiple data sets of the same modality coming from a fixed set of sensors. Cases include:
- Stimuli-data signals recorded using a fixed set of sensors (i.e., a WAMS). These data may include response data to planned perturbations or tests (Hauer et al. 2009).
- System response data to different perturbations, collected by the same set of sensors.
- Historic data collected by a fixed set of sensors or a WAMS.

Case 2. Fusion of multimodal data coming from different sensors. This data may correspond to a single perturbation response, recorded at two or more locations using different sensor technologies or PMU channels.

Case 3. Fusion of homogeneous (unimodal data) data collected by different Phasor Data Concentrator (PDCs) or WAMS that may include different sampling times or different sets of sensors.

Case 4. Fusion of distribution data collected using distribution PMUs.

These cases are discussed separately below. First, the notion of unimodal (single type) data fusion is introduced.

Data fusion approaches can be categorized in a number of ways. Among them, unsupervised data fusion approaches are of special interest for power system applications. Examples include multimodal PLS, multimodal Canonical Correlation Analysis (CCA), Independent Vector Analysis (IVA), joint Independent Component Analysis (ICA), and multiview diffusion maps (DM), to name a few approaches. The theoretical assumptions behind these approaches may not be consistent with the inherent nature of power system data or may not be appropriate for real-time applications. Table 5.1 list some recent approaches developed in the literature.

5.3.2 Unimodal (Single Type) Analysis Approaches

Unimodal or single type data are routinely analyzed or fused for prognosis and decision making. For example, one may wish to use frequency recordings collected on a WAMS to assess power system health, or the extent and distribution of system damage following loss-of-generation events. Separate analyses of different data types measured by the same sensor network, however, may not capture critical associations and potential causal relationships between them.

TABLE 5.1

Multivariate Data Fusion Analysis Techniques

Approach	References
Multiblock PCA	Bakshi (2004), Wold et al. (1996).
Multiblock PLS	Lopes et al. (2002), Krishnan et al. (2011)
Multi-view diffusion maps	Lindenbaum et al. (2015)
Multiple-set canonical correlation analysis	Kettenring (1971)
Independent vector analysis	Kim et al. (2006)
Joint ICA	Calhoun et al. (2006)
Higher-order generalized SVD	Ponnapalli et al. (2011)
Multimodal CCA	Cheveigné et al. (2018), Correa et al. (2008)

Figure 5.1 shows an example of unimodal data collected by different sensors, combined into a single data set of the form (5.2).

Here, each data set is assumed to be of the form

$$X^l = \begin{bmatrix} x_1 & x_2 & \cdots & x_m \end{bmatrix} = \begin{bmatrix} x_1(t_1) & x_2(t_1) & \cdots & x_{m_l}(t_1) \\ x_1(t_2) & x_2(t_2) & \cdots & x_{m_l}(t_2) \\ \vdots & \vdots & \ddots & \vdots \\ x_1(t_N) & x_2(t_N) & \cdots & x_{m_l}(t_N) \end{bmatrix} \in \Re^{N \times m_l}, \quad (5.2)$$

where m_l represents the number of sensors at area l (Figure 5.2).

A reduced model can be obtained by analyzing a representation of the form (5.1). In the more general (and more interesting) case, data must be fused together using a multivariate technique.

5.3.3 Feature-Level Data Fusion

As observed in Chapter 1, there are several ways in which sensor architectures can be implemented. Data fusion can be accomplished at several levels

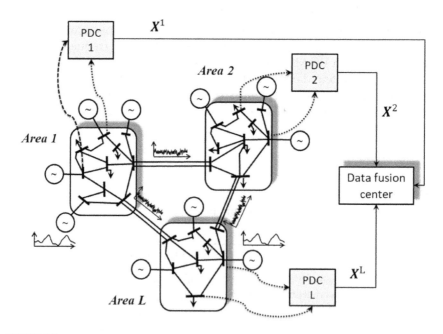

FIGURE 5.2
Schematic illustrating the notion of unimodal data fusion.

TABLE 5.2

Examples of Feature-Level and Raw-Level Data Fusion

Level	Feature
Raw data	Bus voltage magnitudes and phase angles, active/reactive power, speed, frequency, …
Feature level	Modal features, trends, etc.
Decision level	Global signals

(i.e., low-, mid-, and high-levels). At the outset, three application levels of special importance to power system monitoring are considered

1. application to raw data of the same type or modality;
2. application to feature-level data; and
3. application to decision level data.

Table 5.2 lists some examples of potential raw data, feature data, and decision level data. Recent analytical experience suggests that combined raw and feature level data may provide a better visualization and characterization of power system dynamic behavior (Dutta and Overbye 2014).

Following Lafon et al. (2006), the goal is to fuse data by computing a unified low-dimensional representation (a reduced feature space or embedding scheme). In this approach, features extracted from different measurement sets are treated equally and they are often concatenated into a single feature vector for further classification or clustering tasks. An example is modal information extracted from individual data sets. As the representation, distribution, and scale of different data sets may be very different, quite a few studies have suggested limitations to this kind of fusion.

5.4 Multisensor Data Fusion Framework

The analysis of various data sets simultaneously is a problem of growing importance.

These methods have the potential to enhance fundamental understanding of multivariate processes and may prove useful in health diagnosis, monitoring, and the analysis and visualization of power system disturbances (Cabrera et al. 2017).

To introduce these ideas assume that multiple, independent data sets $X^1 \in \mathfrak{R}^{N \times 1}$, $X^2 \in \mathfrak{R}^{N \times m_2}$, ..., $X^L \in \mathfrak{R}^{N \times m_L}$ are collected on different sensors—refer to Figure 5.3. In combination, different data types can provide complementary information about system behavior and capture various time scales.

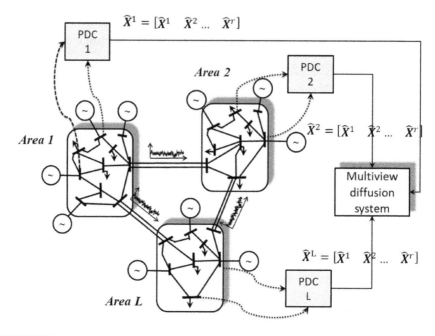

FIGURE 5.3
Multisensor data sets for data fusion analysis.

As discussed in Section 5.3, several practical applications are of interest:

- Data could represent different variables measured at the same set of multichannel PMUs (i.e., power, frequency, bus voltage magnitude, and phase angle, etc.).
- Data measured at various sets of sensors or neighboring utilities.

Next, some generic approaches to data fusion are examined in the context of power system oscillatory monitoring.

5.4.1 Blind Source Separation-Based Data Fusion

These methods have their origin in the solution of mode identification within the framework of Second-Order Blind Source Identification (SOBI). Discussions regarding the application of BSS (blind source separation) to power system data are provided in Messina (2015).

Within the framework of blind source separation (de Cheveigné et al. 2019) each data matrix may be assumed to consist of linear combinations of a set of sources common to all data matrices, plus noise, namely

$$X^l = \begin{bmatrix} x_1 & x_1 \ldots & x_{m_L} \end{bmatrix} = H_l S + N_l \tag{5.3}$$

In this representation, X^l is a matrix of measurements, the H_l are unknown mixing matrices representing the stationary linear transformations from the vector of source signals to the ensemble of observations, and N_l is the matrix of measurement noises. The goal is to determine the common sources S.

For a single set of measurements, the source separation problem can be defined as the simultaneous estimation of the mixing matrix H_l and the underlying oscillatory components. This involves solving two major problems, normalizing data and joint data fusion. In the first step, each matrix is first whitened using PCA, and each principal component is scaled to unit norm. After temporal concatenation of the resulting data sets, a second application of PCA results in a single matrix $Y = [X^1 \quad X^2 \cdots \quad X^L]V$ of dimension $N{\times}m_T$, where $m_T = m_1 + \ldots + m_L$. The submatrices V^l of V define transforms applicable to each data matrix, $Y^l = X^l V^l, l = 1, \ldots, L$. Variations to this approach in the framework of multiblock PCA are discussed in Chapter 9. The procedure is briefly summarized in the Multiway CCA Algorithm box.

Multiway CCA Algorithm

Given multiple sets of observations, $X^1, X^2, \ldots, X^L$:

1. Use PCA to whiten the data matrices, $X^l, l = 1, \ldots, L$ and obtain whitened matrices $\tilde{X}^l$.

2. Concatenate the temporal data sets in step 1 as

$$\tilde{X} = \begin{bmatrix} \tilde{X}^1 & \tilde{X}^2 \cdots & \tilde{X}^l \end{bmatrix}.$$

3. Apply PCA to $\tilde{X}$ and compute

$$Y = \begin{bmatrix} X^1 & X^2 \cdots & X^L \end{bmatrix} V.$$

4. Compute the transformations

$$Y^l = X^l V^l, l = 1, \ldots, L. \tag{5.4}$$

5. Obtain the matrix of summary components, Y, as the sum of components in (5.3).

5.4.2 Feature-Based Fusion of Multimodal Data

As discussed in Sections 5.3.1 and 5.3.2, direct fusion of multimodal data may be difficult. Given the inherent dissimilarity of multimodal data, an alternative approach consists of defining a common set of features belonging to both individual data sets. This may result in improved performance of the data fusion methods and facilitate classification.

For power system monitoring, models that focus on specific features (i.e., oscillatory behavior) are of special interest. Here, a model reduction technique is independently applied to each data matrix X^l (sensor level), resulting in a model representation of the form

$$\widehat{X}^l \approx \sum_{j=0}^{d_l} a_j \psi_j^T, \ l = 1, \ldots, L, \tag{5.5}$$

where d_l is the number of relevant modal components and a_j and ψ_j represent, respectively, the temporal and spatial components associated with mode j. Usually, the trend component $j = 0$, is not included. Typical analysis approaches include wavelet analysis, Hilbert-Huang transform analysis, DMD, and PCA to mention a few.

Using this approach, the modal decomposition of the lth sensor can be expressed as

$$X^l = \begin{bmatrix} x_1 & x_2 & \cdots & x_{m_L} \end{bmatrix} = \begin{bmatrix} x_1(t_1) & x_2(t_1) & \cdots & x_k(t_1) \\ x_1(t_2) & x_2(t_2) & \cdots & x_k(t_2) \\ \vdots & \vdots & \ddots & \vdots \\ x_1(t_N) & x_2(t_N) & \cdots & x_k(t_N) \end{bmatrix} \in \Re^{N \times m_l} \tag{5.6}$$

where, in this case, L denotes the number of modal components at the lth sensor. Combining, the individual sensor models results in a regional (PDC model) of the form

$$\widetilde{X}^l = \begin{bmatrix} \underbrace{x_1 \quad x_2 \quad \cdots \quad x_{m_1}}_{Sensor\ 1} & \underbrace{x_1 \quad x_2 \quad \cdots \quad x_{m_2}}_{Sensor\ 2} & \cdots & \underbrace{x_1 \quad x_2 \quad \cdots \quad x_{mL}}_{Sensor\ m} \end{bmatrix} \tag{5.7}$$

and, for the overall system

$$\widetilde{X} = \begin{bmatrix} \widetilde{X}^1 & \widetilde{X}^2 \ldots & \widetilde{X}^L \end{bmatrix} \tag{5.8}$$

By systematically choosing the frequency range of interest (i.e., the low-frequency inter-area range), a multiband-based data fusion technique can be developed. An illustration of this model is given in Figure 5.4.

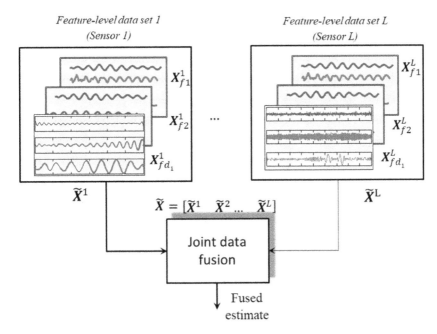

FIGURE 5.4
Feature-level data fusion.

A data fusion technique is then applied to each decomposition of the form (5.5), and the fused models $\hat{Z}_{fl}$ are combined to obtain a low-dimensional representation (reduced feature space)

$$\tilde{X}_f = \left\{ \hat{Z}_{fl} \quad \hat{Z}_{fl} \dots \quad \hat{Z}_{fl} \right\} \tag{5.9}$$

Upon determining a feature-based low dimensional representation, a remaining challenge is to recombine modal characteristics by discarding uninteresting behavior. This approach could be implemented in many ways, including Canonical Correlation Analysis (CCA), Independent Component Analysis (ICA), and PCA. Figures 5.5 and 5.6 show two basic approaches to multiway data fusion.

Other more general representations are discussed in Section 5.5 of this chapter.

Some limitations are inherent to these approaches. First, each data set is analyzed separately. This prevents the analysis of correlations between the data sets.

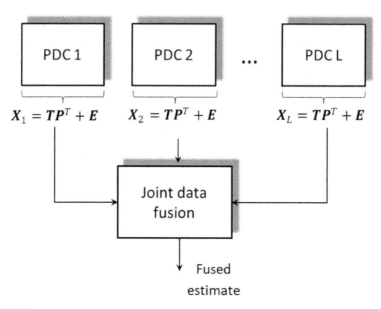

FIGURE 5.5
Feature-level data fusion. CCA interpretation.

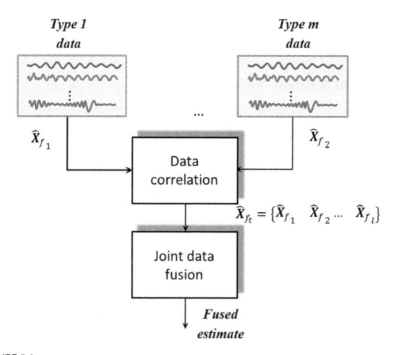

FIGURE 5.6
Illustration of joint data fusion of multimodal data. Case 4 above.

5.5 Multimodal Data Fusion Techniques

Modern approaches to data fusion account for multimodal data (in Section 5.5). In recent years, several multiblock analysis methods with the ability to simultaneously analyze multi-type (multiblock) data have been developed. These include hierarchical PCA, hierarchical PLS, multiway DMs, and consensus PCA, as well as recent extensions such as JIVE (Joint and Individual Variation Explained). Of primary interest are applications to modeling and monitoring dynamic processes.

In this section, several multimodal data fusion techniques are reviewed in the context of power system data fusion.

5.5.1 Multiscale PCA/PLS Methods

Some of the early approaches to dealing with multisensory data involve using PCA. Principal component analysis transforms an $N \times m$ observation matrix, X, by combining the variables as a linear weighted sum (Wold et al. 1996) as

$$X = \widehat{X} + TP^T + E = \widehat{X} + \sum_{i=1}^{a} t_i p_i^T + E \tag{5.10}$$

where $\widehat{X}$ is the center of X, $T \in \mathfrak{R}^{m \times a}$ and $P \in \mathfrak{R}^{a \times N}$ are the scores and loading matrices, respectively. The matrix E contains error terms.

In analyzing this model, it should be noted that:

- The first principal component describes the largest amount of variation in the observation matrix, X.
- The leading vectors are orthonormal and provide the direction with the maximum variability.
- The scores from the different principal components are the coordinates of the objects in the reduced space.

The single-scale PCA analysis can be extended to the multiple data sets using the framework developed in Nomikos and McGregor (1994) and Bakshi (2004). Extensions to this approach include the Joint and Individual Variation Explained (JIVE) method (Lock et al. 2013). See Smilde et al. (2003) for detailed discussion of these methods.

Given a model of the form (5.2), the multiblock PCA model is defined as

$$X = \begin{bmatrix} X_1 & X_2 \dots & X_L \end{bmatrix} = \begin{bmatrix} T_1 P_1^T & T_2 P_2^T \dots & T_L P_L^T \end{bmatrix} \tag{5.11}$$

where

$$T_p = \sum_{i=1}^{K} w_i t_i \tag{5.12}$$

and

$$X_i = T_i P_i^T + \varepsilon_i, \, i = 1, \ldots, L$$

where T_p is the vector of super scores, the T_i denote the block scores, and the P_i are the block loadings: the w_i represent block weights (Xu and Goodacre 2012). They show the relative contribution of each block to the global trend in the data summarized in T_p.

The analysis procedure divides into three principal phases: (a) assembling the individual data sets, X^l, (b) unfolding the data (refer to Figure 5.1), and (c) extracting relationships between measured data from the concatenated model $\widehat{X} = TP^T + E$.

As pointed out by Lock et al (2013), while these methods synthesize information from multiple data types, they do not distinguish between joint or individual effects. Extension of this basic representation to describe patterns of relationships between PDCs and a global data fusion center (super PDC) are described in later chapters. As a by-product, by combining the proposed approach with a modal decomposition technique, feature (modal) level data fusion is possible.

This approach is especially well suited to detect changing patterns between data in different PDCs and detect disturbances on a global scale.

5.5.2 Multiview Diffusion Maps

In Lindenbaum et al. (2015), an approach to cluster multimodal data was proposed. Given multiple sets of observations, $X^1, X^2, \ldots, X^L$, the underlying idea is to seek for a lower dimensional representations that achieves two objectives: (a) Preserve the interactions between multidimensional data points within a given view X^2 and (b) preserve the interactions between the data sets $X^1, X^2, \ldots, X^L$.

According to this idea, the multiview kernel can be expressed as

$$\widehat{K} = \begin{bmatrix} 0_{M \times M} & K^1 K^2 & K^1 K^3 \cdots & K^1 K^p \\ K^2 K^1 & 0_{M \times M} & K^2 K^3 \cdots & K^2 K^p \\ K^3 K^1 & K^3 K^2 & 0_{M \times M} \cdots & K^3 K^p \\ \vdots & \vdots & \ddots & \vdots \\ K^p K^1 & K^p K^2 & K^p K^3 \cdots & 0_{M \times M} \end{bmatrix} \in \Re^{l \times m \times l \times m} \tag{5.13}$$

Once a model of the form (5.13) is obtained, the conventional DM approach can be applied to fuse multisource data. It is apparent from (5.13) that a costly part of these algorithm is the construction of the feature matrices K^l.

A simple algorithm to compute a multiview representation is outlined in the Multiview Diffusion Algorithm box:

Multiview Diffusion Algorithm

Given multiple sets of observations, $X^1, X^2, \ldots, X^L$:
6. Compute the local affinity matrices A^l and the Gaussian kernel K^l for each data set (view) using (3.26).
7. Select the appropriate local $\varepsilon_i, \varepsilon_j$ and construct the normalized affine matrices.
8. Compute the multiview kernel using (3.27).
9. Normalize the kernel

$$\hat{P} = \hat{D}^{-1}\hat{K}. \tag{5.14}$$

10. Select the appropriate global scales $\varepsilon_i, \varepsilon_j$ by using the approach in Arvizu and Messina (2016).
11. Calculate the spectral decomposition of $\hat{P}$ and construct the low-dimensional mapping

$$\hat{\Psi}(\hat{X}) = \left[\hat{\Psi}(\hat{X}^1) \quad \hat{\Psi}(\hat{X}^2) \ldots \quad \hat{\Psi}. \right.$$

12. Determine the temporal evolution of the multiview model.

Remarks:

- Key to the above representation is the appropriate calculation of global scales $\varepsilon_i, \varepsilon_j$.
- An alternative representation of Multiview diffusion maps can be obtained from multiset canonical correlation analysis, through the optimization of an objective function of a correlation matrix of the canonical variables (Li et al. 2009).

Given a set of data matrices, $X^1, X^2, \ldots, X^L$, the objective function for maximization of the overall correlation among canonical variables is defined as

$$\max_{w_1, w_2, \ldots, w_N} \rho = \sum_{i \neq j}^{N} w_i^T X_i X_j^T w_j$$

$$\text{s.t.} \quad \frac{1}{N} \sum_{i=1}^{N} w_i^T X_i X_j^T w_j = 1 \tag{5.15}$$

It can be shown that the optimization problem can be transformed into the generalized eigenvalue problem $(\mathbf{R} - \mathbf{S})w = \rho S w$, where

$$R = \begin{bmatrix} X^1(X^1)^T & X^1(X^2)^T & X^1(X^3)^T & \cdots & X^1(X^N)^T \\ X^2(X^1)^T & X^2(X^2)^T & X^2(X^3)^T & \cdots & X^2(X^N)^T \\ X^3(X^1)^T & X^1(X^1)^T & X^3(X^3)^T & \cdots & X^3(X^N)^T \\ \vdots & & \vdots & \ddots & \vdots \\ X^N(X^1)^T & X^N(X^2)^T & X^N(X^3)^T & \cdots & X^N(X^N)^T \end{bmatrix}$$

$$S = \begin{bmatrix} X^1(X^1)^T & & & \\ & X^2(X^2)^T & & \\ & & \ddots & \\ & & & X^N(X^N)^T \end{bmatrix}$$

$$w = \begin{bmatrix} w_1 & w_2 \ldots & w_N \end{bmatrix}$$

By combining these ideas with Lindenbaum's approach, other new formulations can be obtained. This is a subject that warrants further investigation. Variations to these approaches involving tensor-based representations are presented in Zhang et al. (2011).

5.5.3 Partial Least Squares

Another class of methods to fuse (regress) data are based on the notion of partial least squares correlation (Adbi and Williams 2013, Van Roon et al. 2014). As the simplest example, consider two data sets $X \in \mathfrak{R}^{N \times m_1}$ and $Y \in \mathfrak{R}^{N \times m_2}$, which in the more general case are assumed to have different types (and therefore mixed units). In the present context they could represent measurements such as voltage, frequency, reactive power, etc., and are assumed to have the same length N and have a different number of sensors, $m_1 \neq m_2$. The relationship between these two centered sets of variables is given by the covariance matrix $R = Y^T X$.

Conceptually, these techniques decompose the data into a set of statistically independent and orthogonal components. Each component is defined in time by a unique series of weights assigned to each of the time samples; these weights are called *component loadings*. As discussed in Chapter 4, the extent to which each principal component contributes to the particular waveform in the original data set is reported as a *component score* (Rosipal et al. 2005).

From the preceding results in Section 5.3.2, it follows that

$$X = TP^T + E$$
$$Y = UQ^T + F \tag{5.16}$$

and

$$U = TD$$

where T and U, with dimension nxa are X-scores and Y-scores; P and P are the X-loadings and Y-loadings, respectively; E and F are X-residuals and Y-residuals, respectively and D is a diagonal matrix w_i.

The cross-block correlation matrix R can be computed as

$$R = (Y)^T X \tag{5.17}$$

Singular value decomposition of R yields

$$R = U\Sigma V \tag{5.18}$$

with the latent variables $L_x = XV$ and $L_y = YU$.

A physical interpretation of the model is then possible:

- Matrices U_x and U_y give the data sets spatial patterns or shapes.
- The correlation matrix R gives the correlation between a given output signal, Q_j, and a given input signal, V_k.

5.5.4 Other Approaches

Several other extensions of the above strategies have been proposed in the literature and are surveyed by Smilde et al. (2003). These include joint blind source separation by multiset canonical correlation analysis (Li et al. 2009).

5.6 Case Study

In this example, multivariate cluster analysis techniques were applied to simulation data from the Australian test system in Figure 3.10. The goal is to characterize critical associations or patterns between bus voltage magnitudes and reactive power outputs from generators and SVCs. Attention is focused on two main aspects, namely, the identification of phase relationships between reactive power and voltage signals and the computation of voltage to reactive power sensitivities to system perturbations.

For purposes of illustration, 18 reactive power output signals, including reactive power at 14 generators and 4 SVCs, and 59 bus voltage magnitudes were selected for analysis. Figure 5.7 shows some selected measurements, while Table 5.1 summarizes the main characteristics of the sets. Insight into the nature of phase relationships is given in Figure 5.8 that plots the behavior of coherent signals.

The observational data is defined as

$$X_V = \begin{bmatrix} V_1 & V_2 \dots & V_{59} \end{bmatrix} \in \mathfrak{R}^{2401 \times 59} \tag{5.19}$$

$$X_Q = \begin{bmatrix} X_{Qg} & X_{QSVC} \end{bmatrix} = \begin{bmatrix} Q_1 & Q_2 \dots & Q_{19} \end{bmatrix} \in \mathfrak{R}^{2401 \times 18} \tag{5.20}$$

where $V_j(t)$, $j = 1, \dots, 59$ denotes bus voltage magnitudes and $Q_j(t)$, $j = 1, \dots, 18$ are the output reactive power signals from synchronous generators and SVCs. The description of the data sets is given in Table 5.3.

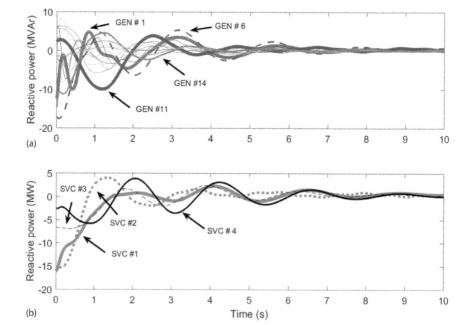

FIGURE 5.7
Selected time series following a short circuit at bus 217. (a) Generators reactive power output and (b) SVC reactive power output.

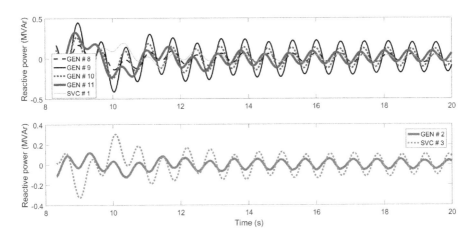

FIGURE 5.8
Detail of system response.

TABLE 5.3

Selected System Measurements

Data Subtype	Features
Bus voltage magnitudes (X_V)	59 signals
Generator reactive power output signals (X_{Qg})	15 generator reactive power output signals (Gens: 101, 201, 202, 203, 204, 302, 302, 303, 401, 402, 403, 404)
SVC reactive power output signals	5 SVC reactive power output signals (SVCs: ASVC_2, RSVC_3, BSVC_4, PSVC_5, SSVC_5 in Figure 3.10)

Inspection of Figures 5.7 and 5.8 reveals that:

- Generators 8 through 11 and SVC # 1 swing in phase.
- Further, generator # 2 and SVC # 3 swing in phase.

Partial least squares analysis provides a natural way to explore associations between individual data blocks or variables. As a first step to evaluate the applicability of these models each data set was analyzed independently. For reference, Figure 5.9 shows the spatial patterns extracted from X_V and X_Q in equations (5.19) and (5.20) analyzed separately and obtained using PCA.

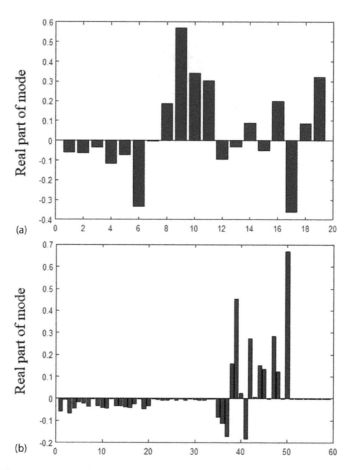

FIGURE 5.9
Real part of PCA-based reactive power (a) and bus voltage magnitude (b) modes.

5.6.1 Partial Least Squares Correlation

The approach in Section 5.6.2 was used to determine relationships between voltage and output reactive power from SVCs and generators. The primary goal is to determine patterns of behavior of interest for voltage control.

As a first step, voltage and reactive power data are column centered as

$$
\begin{aligned}
\widehat{X}_V &= X_V - 1_m x_{mean_v}^T \\
\widehat{X}_Q &= X_Q - 1_m x_{mean_q}^T
\end{aligned}
\tag{5.21}
$$

The correlation matrix is then computed as

$$R = \left(\widehat{X}_Q\right)^T \widehat{X}_V \in \mathfrak{R}^{59 \times 19} \tag{5.22}$$

Each row of matrix R, gives the correlation between a given reactive power output, Q_j, and a given bus voltage magnitude, V_k. Direct calculation of the covariance matrix, however, may result in numerical errors, especially for large systems and is sensitive to scaling. Figure 5.10 shows the right eigenvector plot U_{xi} and U_{yj}. Correlation analysis that SVCs # 3 and 4 swing in opposition to SVCs # 1, 2, and 5.

Also of interest, Table 5.4 shows correlation measures from (5.22). This is consistent with the physical understanding of the problem.

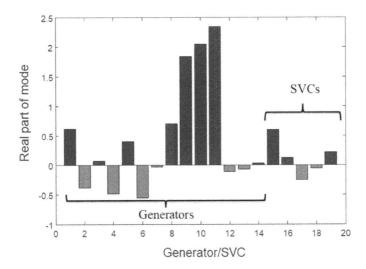

FIGURE 5.10
Partial least squares correlation analysis of reactive power time series.

TABLE 5.4

Spatial Correlation between SVC Reactive Power
Output and Bus Voltage Magnitudes

SVC Bus	Dominant Bus Voltage Magnitudes
205	209, 212, 208, 414, 202
313	314, 304, 307, 306
412	403, 411, 414
507	509, 505, 305
509	509, 505, 315, 207, 203

5.6.2 Multiblock PCA Analysis of Measured Data

Further insight into the nature of dynamic patterns can be obtained from multiblock (joint) analysis of the observational matrices X_V, X_Q and X_{SVC} in Figures 5.7 and 5.8. In this case, the concatenated data is expressed in the form

$$X = \begin{bmatrix} X_V & X_{Qg} & X_{QSVC} \end{bmatrix} \tag{5.23}$$

Because of the different physical units, separate pre-processing methods were required for the three different types of data. Figure 5.11 shows the shape extracted from multiblock PCA analysis of the multivariate set in (5.19) and (5.20). Results are found to be in good agreement with Figures 5.9b and 5.10.

Figure 5.12 compares the frequency-based shape extracted from 19 frequency measurements extracted using DMs and Laplacian eigenmaps. Results are found to be consistent although some differences are noted.

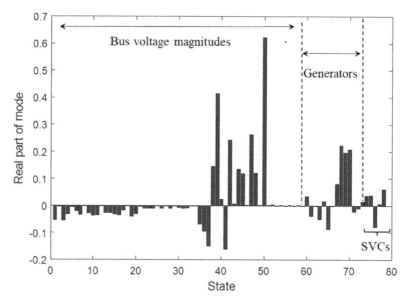

FIGURE 5.11
Schematic illustration of dimensionality reduction.

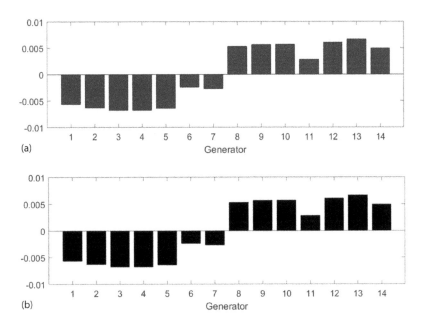

FIGURE 5.12
(a) Diffusion maps and (b) Laplacian eigenmaps.

References

Abdi, H., Williams, L. J., Partial least squares methods: Partial least squares correlation and partial least square regression, in Reisfeld, B., Mayeno, A. N., (Eds.), *Computational Toxicology, Volume II, Methods in Molecular Biology*, 930, pp. 543–579, Human Press-Springer, New York, 2013.

Abdulhafiz, W. A., Khamis, A., Handling data uncertainty and inconsistency using multisensor data fusion, *Advances in Artificial Intelligence*, 241260, 1–11, 2013.

Allerton, D. J., Jia, H., A review of multisensor fusion methodologies for aircraft navigation systems, *The Journal of Navigation*, 58(3), 405–417, 2005.

Arvizu, C. M. C., Messina, A. R., Dimensionality reduction in transient simulations: A diffusion maps approach, *IEEE Transactions on Power Delivery*, 31(5), 2379–2389, 2016.

Bakshi, B. R., Multiscale PCA with application to multivariate statistical process monitoring, *AIChE Journal*, 44(7), 1596–1610, 2004.

Cabrera, I. R., Barocio, E., Betancourt, R. J., Messina, A. R., A semi-distributed energy-based framework for the analysis and visualization of power system disturbances, *Electric Power Systems Research*, 143, 339–346, 2017.

Calhoun, V. D., Adali, T., Pearlson, G. D., Kiehl, K. A., Neuronal chronometry of target detection: Fusion of hemodynamic and event-related potential data, *NeuroImage*, 30, 544–553, 2006.

Claudia M. C. Arvizu, A. R. Messina, Dimensionality Reduction in Transient Simulations: A Diffusion Maps Approach, 31(5), IEEE Trans. on Power Delivery, October 2016, pp. 2379–2389.

Correa, N. M., Li, Y. O., Adali, T., Calhoun, V. D., Canonical correlation analysis for feature-based fusion of biomedical imaging modalities and its application to detection of associative networks in schizophrenia, *IEEE Journal of Selected Topics in Signal Processing*, 2(6), 998–1007, 2008.

Dalla Mura, M., Prasad, S., Pacifici, F., Gamba, P., Benediktsson, J. A., Challenges and opportunities of multimodality and data fusion in remote sensing, *Proceedings of the IEEE*, 103(9), 1585–1601, 2015.

de Cheveigné, A., Di Liberto, G. M., Arzounian D., Wong, D. D. E., Hjortkjær, J., Fuglsang, S., Parra, L. C., Multiway canonical correlation analysis of brain data, *NeuroImage*, 186, 728–740, 2019.

Dutta, S., Overbye, T. J., Feature extraction and visualization of power system transient stability results, *IEEE Transactions on Power Systems*, 29(2), 966–973, 2014.

Fusco, F., Tirupathi, S., Gormally, R., Power system data fusion based on belief propagation, *2017 IEEE PES Innovative Smart Grid Technologies Conference Europe (ISGT-Europe)*, Torino, Italy, September 2017.

Ghamisi, P., Rasti, B., Yohoka, N., Wang, Q., Hofle, B., Bruzzone, L., Bovolo, F., Chi, M., Anders, K., Gloaguen, R., Atkinson, P. M., Benediktsson, J. A., Multisource and multitemporal data fusion in remote sensing – A comprehensive review of the state of the art, *IEEE Geoscience and Remote Sensing Magazine*, 7(1), 6–39, 2019.

Hall, D. L., Llinas, J., Multisensor data fusion, *Proceedings of the IEEE*, 85(1), 6–23, 1997.

Hauer, J. F., Mittelstadt, W. A., Martin, K. E., Burns, J. W., Lee, H., Pierre, J. W., Trudnowski, D. J., Use of the WECC WAMS in wide-area probing tests for validation of system performance and modeling, *IEEE Transactions on Power Systems*, 24(1), 250–257, 2009.

Kettenring, J. R., Canonical analysis of several sets of variables. *Biometrika*, 58(3), 433–451, 1971.

Kezunovic, M., Abur, A., Merging the temporal and spatial aspects of data and information for improved power system monitoring applications, *Proceedings of the IEEE*, 93(11), 1909–1919, 2005.

Kim, T., Lee, I., Lee T. W., Independent vector analysis: Definition and algorithms, *Proceedings of the 40th Asilomar Conference on Signals, Systems, and Computers, 1393–1396*, Pacific Grove, 2006.

Krishnan, A., Williams, L. J., McIntosh, A. R., Abdi, H., Partial Least Squares (PLS) methods for neuroimaging: A tutorial and review, *NeuroImage*, 56, 455–475, 2011.

Lafon, S., Keller, Y., Coifman, R. R., Data fusion and multicue data matching by diffusion maps, *IEEE Transactions on Pattern Analysis and Machine Intelligence*, 28(11), 1784–1797, 2006.

Li, Y. O., Adali, T., Wang, W., Calhoun, V., Joint blind source separation by multiset canonical correlation analysis, *IEEE Transactions on Signal Processing*, 57(10), 3918–3929, 2009.

Lindenbaum, O., Yeredor, A., Salhov, M., Averbuch, A., Multiview diffusion maps, *Information Fusion*, arXiv:1508:05550, 2015.

Lock, E. F., Hoadley, K. A., Marron, J. S., Nobel, A. B., Joint and individual variation explained (JIVE) of integrated analysis of multiple datasets, *The Annals of Applied Statistics*, 7(1), 532–542, 2013.

Lopes, J. A., Menezes, J. C., Westerhuis, J. A., Smilde, A. K., Multiblock PLS analysis of an industrial pharmaceutical process, *Biotechnology and Bioengineering*, 80, 419–427, 2002.

Messina, A. R., *Wide-Area Monitoring of Interconnected Power Systems*, IET Power and Energy Series 77, London, UK, 2015.

Nomikos, P., MacGregor, J. F., Monitoring batch processes using multiway principal component analysis, *AIChE Journal*, 40(8), 1361–1375, 1994.

Ponnapalli, S. P., Saunders, M. A., Van Loan, C. F., Alter, O., A higher-order generalized singular value decomposition for comparison of global mRNA expression from multiple organisms, *Plos One*, 6(12), 1–11, 2011.

Rosipal, R., Kramer, N., Overview and recent advances in partial least squares, in Saunders, C., Grobelnik, M., Gunn, S., Shawe-Taylor, J. (Eds.), *Subspace, Latent Structure and Feature Selection, Lecture Notes in Computer Science*, pp. 34–51, Springer, Berlin, Germany, 2005.

Savopol, F., Armenakis, C. Merging of heterogeneous data for emergency mapping: Data integration or data fusion? *Symposium on Geospatial Theory, Processing and Applications*, Ottawa, 2002.

Schmitt, M., Zhu, X. X., Data fusion and remote sensing, *IEEE Geoscience and Remote Sensing Magazine*, 4(4), 6–23, 2016.

Smilde, A. K., Jansen, J. J., De Jong, S., A framework for sequential multiblock component models, *Journal of Chemometrics*, 17, 323–337, 2003.

Van Roon, P., Zakizadeh, J., Chartier, S., Partial Least Squares tutorial for analyzing neuroimaging data, *Tutorials in Quantitative Methods for Psychology*, 10, 200–215, 2014.

Williams, M. O., Rowley, C. W., Mezic, I., Kevrekidis, I. G., Data fusion via intrinsic dynamic variables: An application of data-driven Koopman spectral analysis, *EPL (European Physics Letters)*, 109(4), 1–6, 2015.

Wold, S., Kettaneh, N., Tjessem, K., Hierarchical multiblock PLS and PC models for easier model interpretation and as an alternative to variable selection, *Journal of Chemometrics*, 10, 463–482, 1996.

Worden, K., Staszewski, W. J., Hensman, J. J., Neural computing for mechanical system research, *Mechanical Systems and Signal Processing*, 25, 4–111, 2011.

Xu, Y., Goodacre, R., Multiblock principal component analysis: An efficient tool for analyzing metabolomics data which contain two influential factors, *Metabolomics*, 8(S1), 37–51, 2012.

Zhang, J., Vittal, V., Sauer, P., Networked Information Gathering and Fusion of PMU Data, Future Grid Initiative White Paper, Power Systems Engineering Research Center, PSERC Publication 12-07, May 2012.

Zhang, Y., Zhou, G., Zhao, Q., Onishi, A., Jin, J., Wang, X., Cichocki, A., Multiway canonical correlation analysis for frequency component recognition in SSVEP-based BCIs, in Lu, B. L., Kwok, J. (Eds.), *International Conference on Neural Information Processing*, Vol. 7062, pp. 287–295, Springer, Berlin, Germany, 2011.

6

Dimensionality Reduction and Feature Extraction and Classification

6.1 Introduction: Background and Driving Forces

Dimensionality reduction has been an active field of research in power system analysis and control in recent years. These methods produce low-dimensional representations of high-dimensional data, where the representation is chosen to accurately preserve relevant structure. A survey of dimension reduction techniques is given in (Fodor 2002).

Dimensionality reduction is important for various reasons. First, mapping high dimensional data into a low-dimensional manifold facilitates better understanding of the underlying processes and aids to eliminate or reduce irrelevant features and noise as well as to facilitate retrieval of selected dynamic information. In addition, an effective reduced description provides some intuition about the physical system or process and allows visualization of the data using two or three dimensions (Mendoza-Schcrock et al. 2012). Finally, with a successful application of these methods, it is possible to find features that may not be apparent in the original space.

Most approaches to dimensionality reduction are based on three main steps: (a) Data normalization, (b) Dimension reduction, and (c) Clustering and data classification.

Multivariate dimension reduction techniques have been applied to power system measured data including POD/PCA analysis, PLS and their nonlinear variants, DM, and DMD to mention a few approaches (Chen et al. 2013; Ramos et al. 2019; Arvizu and Messina 2016). These models have been shown to be useful in the analysis and characterization of the global behavior of transient processes, as well as to extract and isolate the most dominant modes for model reduction of complex systems. Current analysis approaches, however, are limited in that they consider that the distribution of physically relevant states is highly clustered around a set of much lower dimensionality, and often result in full matrix representations.

In this chapter, a brief description and background of dimensionality reduction is presented, including a general review of related work in this field of research. The context and framework for the development of reduced order representations are set forth, with a particular focus on the extraction of spatiotemporal patterns of large data sets.

6.2 Fundamentals of Dimensionality Reduction

6.2.1 Formal Framework

Dimensionality reduction techniques map (project) high-dimensional data points in the original observation space $(X^{m \times D})$ to a low-dimensional space $(\widehat{X}^{m \times d})$ by constructing meaningful coordinates that are combinations of the original feature vectors. Usually, this is achieved by optimizing certain criterion or objective function—see Van der Maaten et al. (2009), for a taxonomy of dimensionality reduction methods and the definition of constraints.

In practice, the number of coordinates $d \ll D$ is selected to retain or preserve as much information as possible; in power system applications, m typically represents the number of sensors (or network points or states) and d is the number of coordinates, which are often associated to global system behavior.

Figure 6.1 schematically summarizes the context and main purposes of dimensionality reduction. Inputs to this model are raw data collected from dynamic sensors or selected features derived from measured data. With current multichannel PMUs and other specialized sensors coexisting together,

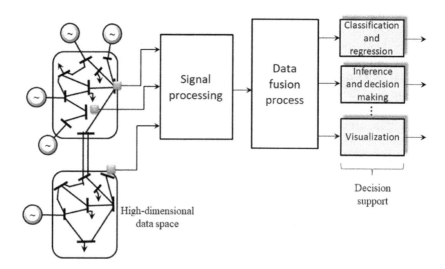

FIGURE 6.1
Basic dimensionality reduction pipeline.

inputs to the data fusion process are multiple sets of data often differing in dimensionality, nature, and even resolution.

Emerging from the data fusion center are techniques to visualize, classify, correlate, and forecast system dynamic behavior. Analysis tasks such as clustering, classification and regression, visualization, and event or anomaly detection may be carried out efficiently in the constructed low dimensional space (Chen et al. 2013).

Three main advantages can be obtained using such an approach: (a) Hidden parameters may be discovered, (b) Noise effects can be suppressed or reduced, and (c) Interesting and significant structures (patterns) are revealed or highlighted such as change points and hidden states. In addition, data can be efficiently visualized using a few dimensions.

The application of common dimensionality reduction techniques, however, faces several challenges:

- Systematically choosing effective low-dimensional coordinates is a difficult problem. Usually the mapping is non-invertible and special techniques are needed to relate the observed behavior in the low-dimensional space to the original (data) space (Arvizu and Messina 2016).
- Linear analysis methods may fail to characterize nonlinear relationships in measured data and result in poor or incorrect classification or clustering of measured data.
- Dimensionality reduction methods in other fields are often developed for data that have no dynamics (Berry et al. 2013). Moreover, many power system dynamic processes can be characterized by time scale separation which may affect the performance of many techniques.

The following sections examine some open issues in the application of linear and nonlinear dimensionality reduction techniques to measured power system data in the context of data fusion and data mining applications.

6.2.2 Numerical Considerations

The realization of practical dimensionality reduction techniques is challenging due to the large number of measurements collected, noise, and other effects. By reducing the model to the number of sensors or network points (m), the size of the problem can be substantially reduced. This task is also referred to as *input space reduction*.

With the increasing size of recording lengths (N) with ($N > m$), algorithms with the ability to compress system information are needed. The issue of near real-time processing (N small), on the other hand, requires special analysis techniques.

Spectral dimensionality reduction algorithms perform eigen decomposition on a matrix that is proportional in size to the number of data points;

when the number of data points increases, so does the time required to compute the eigen decomposition. For very large data sets, this decomposition step becomes intractable.

For typical power system applications involving WAMS, the dimension m is moderately large (in the order of a few dozen to hundreds of points or sensors), and therefore, efficient data reduction techniques can be developed.

6.3 Data-Driven Feature Extraction Procedures

6.3.1 Feature Extraction

Feature extraction consists of finding a set of measurements or a block of information with the objective of describing in a clear way the data or an event present in a signal. These measurements or features can then be used for detection, classification, or regression (prediction) tasks in power system analysis. Feature selection is also important for model reduction of power grid networks and visual representation of monitored system locations (Lee 2018; Ghojogh et al. 2019). As noted by Lunga et al. (2014), however, feature selection and feature extraction may result in some loss of information relative to the original (unreduced) data.

Formally, given a measurement matrix, $X = \begin{bmatrix} x_1 & x_2 & \cdots & x_N \end{bmatrix} \in \mathfrak{R}^{m \times N}$, feature extraction aims at determining a reduced order representation of the data $\widehat{X} = \begin{bmatrix} \hat{x}_1 & \hat{x}_2 & \cdots & \hat{x}_n \end{bmatrix} \in \mathfrak{R}^{p \times N}$, such that $p \leq m$. Usually, the quality of the low-dimensional representation can be assessed using techniques such as the residual variance (Tenenbaum et al. 2000; Dzemyda et al. 2013; Lu et al. 2014).

6.3.2 Feature Selection

Feature selection aims at selecting a subset of the most representative features $\{\hat{x}_1 \quad \hat{x}_2 \quad \cdots \quad \hat{x}_p\} \subseteq \{x_1 \quad x_2 \quad \cdots \quad x_m\}$ according to an objective function or criteria (Yan et al. 2006). At its most basic level, feature selection aims at determining a model of the form $\widehat{X} = WX$, where W is usually a binary matrix.

The reconstruction error can be expressed for the ith signal in the form

$$\epsilon_i = abs\|x_i - \check{x}_i\|, i = 1, \ldots, n \tag{6.1}$$

or, at a global level, as

$$\epsilon = abs\|X - \widehat{X}\|. \tag{6.2}$$

Feature extraction analysis for power system applications differs from that in other fields in many respects. First, features of interest involve both pre-fault and post-fault system conditions (Li et al. 2015). Further, power system data is time-ordered, noisy, and uncertain.

6.4 Dimensionality Reduction Methods

Dimensionality reduction methods produce low-dimensional representations of high-dimensional data where the representation is chosen to preserve or highlight features of interest according to a suitable optimization criterion that is specific to each method—refer to Lunga et al. (2014) for a description of affinities and constraints for common embedding algorithms. Dimensionality reduction techniques are often divided into convex and non-convex according to the optimization criteria utilized (Van der Matten et al. 2006). Other criteria are proposed in (Lee and Verleysen 2007).

Nonlinear dimensionality reduction methods can be broadly characterized as selection-based and projection-based. Among these methods, spectral dimensionality reduction (manifold-based) techniques have been successfully applied to power system data. Other dimensionality reduction tools, such as Koopman-based techniques and Markov algorithms are rapidly gaining acceptance in the power system literature (Ramos and Kutz 2019; Messina 2015).

As discussed in Chapters 3 and 4, the primary goal of nonlinear dimensionality reduction techniques is to obtain an optimal low-dimensional basis, $\Psi_j(x)$, for representing an ensemble of high-dimensional data, X. This basis can then be used to formulate reduced-order models of the form

$$\widehat{X} = \hat{x}_1 + \hat{x}_2 \ldots + \hat{x}_d = a_o(t)\psi_o^T(x) + \sum_{j=1}^{m} a_j(t)\psi_j^T(x), \qquad (6.3)$$

where $a_j(t)$ represent temporal trajectories, $\Psi_j(x)$ are the spatial patterns, and $a_o(t)\psi_o^T(x)$ indicates average behavior (Long et al. 2016). Each temporal coefficient, $a_j(t)$, provides information about a particular frequency or a frequency range, while $\Psi_j(x)$ gives information about spatial structure. Such a parameterization allows a broad range of physical processes and models to be studied.

6.4.1 Projection-Based Methods

The underlying idea behind these methods is the spectral decomposition of a (square) symmetric feature matrix (Strange and Zwiggelaar 2014, Singer 2009). Common examples include linear models such as PCA/POD, PLS, nonlinear projection methods, and Koopman-based representations.

To pursue these ideas further, consider a set of points, $X = \{x_i\}_{i=1}^{N} \in \mathfrak{R}^D$ (D-dimensional feature or input space), representing a data matrix in a high-dimensional space $\mathfrak{R}^D$, where D usually corresponds to the number of recorded responses. The goal of dimensionality reduction techniques is to recover a set of d-dimensional data $\widehat{X} = \{\hat{x}_i\}_{i=1}^N \in \mathfrak{R}^d$, with $d \leq D$ such that Y accurately captures the relevant features of X. Conceptually, this is achieved by computing the eigenvectors of a feature matrix derived from X. Implicit in this notion is the existence of a mapping $f \rightarrow \mathfrak{R}^d : \mathfrak{R}^D$ that maps high dimensional data x_i to low-dimensional $\hat{x}_1$.

Regardless of the dimensionality reduction technique adopted, these methods solve an eigenvalue problem of the form

$$P\Psi = \lambda\Psi, \tag{6.4}$$

or, alternatively,

$$L\Psi = \lambda M\Psi. \tag{6.5}$$

In the first case, P represents a Markov or transition matrix, a Hessian matrix, or an alignment matrix to mention a few alternatives. In the second case, L denotes, for instance, the graph Laplacian.

The corresponding spatial structure is given by a set of coordinates of the form

$$G(\Psi) = \{\psi_1, \psi_2, \ldots, \psi_d\} = \left\{ \begin{bmatrix} \psi_{11}^{(1)} \\ \psi_{21}^{(1)} \\ \vdots \\ \psi_{m1}^{(1)} \end{bmatrix}, \begin{bmatrix} \psi_{11}^{(2)} \\ \psi_{21}^{(2)} \\ \vdots \\ \psi_{m1}^{(2)} \end{bmatrix}, \ldots, \begin{bmatrix} \psi_{11}^{(d)} \\ \psi_{21}^{(d)} \\ \vdots \\ \psi_{m1}^{(d)} \end{bmatrix} \right\},$$

which are expected to capture the intrinsic degrees of freedom (latent variables) of the data set. A schematic representation of the reduction process is given in Figure 6.2. Examples of these methods are given Chapters 3 and 5.

Remarks:

- The eigenvectors (diffusion coordinates) ψ_k give a good description of the slow (collective variables) dynamics of the system. Experience shows that only a few coordinates, d, are needed to capture system behavior.

- The mapping $\Psi_d : \mathfrak{R}^D \rightarrow \mathfrak{R}^d$ is only given at the recorded states. Techniques to extend this approach to nearby points in the original space, without requiring full recomputation of a new matrix and its eigenvectors are given in Erban et al. (2007).

- Using this information, the modes that dominate the system response and the global measurements that contribute most to the oscillations can be accurately captured.

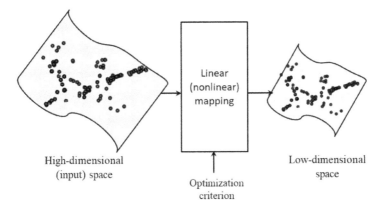

FIGURE 6.2
Illustration of dimension reduction. Dimensionality reduction involves mapping high-dimensional inputs into a low-dimensional representation space that best preserves the relevant features of the data.

Depending on the analysis approach, the modal coordinates, ψ_k, can be orthogonal or non-orthogonal and may exhibit some degree of correlation. This also applies to temporal components in Equation (6.3).

Now, attention is turned the use of dimensionality reduction for cluster validation and classification. Whenever possible, a physical interpretation is provided.

6.4.2 Coarse Graining of Dynamic Trajectories

As discussed earlier, power system data is time-varying in nature and may exhibit time scale separation. Slow motion is usually captured by the first few eigenvectors in Equation (6.5) and the associated temporal coefficients. Identifying these components, however, is not straightforward and may require special algorithms or analysis techniques and some fine tuning.

Two problems are of interest here:

1. Extracting the essential coordinates, d, from the low-dimensional representation, and

2. The analysis of control or physical interactions between states or physical phenomena.

The first problem has been recently studied using diffusion maps (Arvizu and Messina 2016). The second problem in terms of Equation (6.5) concerns the analysis mode interaction described in the context of the perturbation theory of linear systems and is illustrated schematically in Figure 6.3.

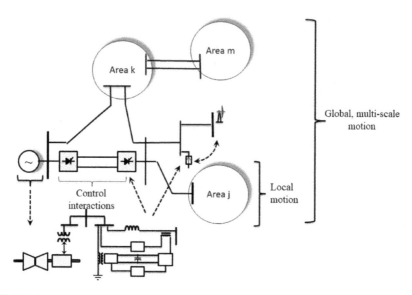

FIGURE 6.3
Schematic illustration of control interactions. (Based on Arvizu and Messina, 2016.)

One way to achieve this is to analyze each component $\widehat{X}_j = a_j(t)\psi_j^T(x)$ separately to infer spatiotemporal content. Another approach is to use a Markov chain analysis.

In Sections 6.4.3 and 6.6.3, two alternatives to dimensionality reduction are explored: Koopman analysis and Markov chains.

6.4.3 Koopman and DMD Analysis

Koopman mode analysis and its variants provide an effective alternative to reduce system dimensionality. In the case of manifold learning techniques, the dominant eigenvectors (coordinates) approximate the most slowly evolving collective modes or variables over the data. Applications of this approach to both simulated and measured data are described in Barocio et al. (2015) and Ramos and Kutz (2019).

From dynamic decomposition theory, these models can be expressed in the general form

$$\widehat{X} = \sum_{j=1}^{p} \phi_j \lambda_j a_j(t) = \widehat{X}_{dmd_1} + \widehat{X}_{dmd_2} + \ldots + \widehat{X}_{dmd_p}, \tag{6.6}$$

where the $\widehat{X}_{dmd_k}$, $k = 1, \ldots, p$ matrices in Equation (6.6) capture specific behavior associated with modes of motion.

6.4.4 Physical Interpretation

The Koopman modes have a useful and interesting interpretation as columns of a matrix of observability measures (Ortega and Messina 2017).

To illustrate these ideas, consider the linear system

$$\dot{z}(t) = Az(t) + Bu(t), \quad \text{with} \quad z(0) = z_o$$
$$y(t) = \hat{C}z(t) \tag{6.7}$$

where $z(t) \in \mathfrak{R}^n$ is the vector of state variables; $y(t) \in \mathfrak{R}^m$ is the vector of system outputs; $A, B,$ and C are constant matrices of appropriate dimensions; and z_o is the vector of initial conditions.

As suggested by Chan (1984), a useful measure of observability is motivated by the expression

$$y(t) = \sum_{i=1}^{n} \hat{C}v_i \left[w_i^T z_o + w_i^T B \int_0^t e^{-\lambda_i t} u(\tau) d\tau \right] e^{-\lambda_i t}, \tag{6.8}$$

where λ_i is the ith eigenvalue, and v_i and w_i are the corresponding right and left eigenvectors.

The observability matrix is defined as

$$\hat{C}V = \begin{bmatrix} \hat{c}_1^T v_1 & \hat{c}_1^T v_1 & \cdots & \hat{c}_1^T v_1 \\ \vdots & \vdots & \ddots & \vdots \\ \hat{c}_1^T v_1 & \hat{c}_1^T v_1 & \cdots & \hat{c}_1^T v_1 \end{bmatrix}. \tag{6.9}$$

It can be proved that, for the case of a linear system $x_{k+1} = Ax_k$, the Koopman eigenfunctions can be defined as $\varphi_j(x_k) = \langle x_k, w_j \rangle$ for the observables $f(x_k) = x_k$.

It has been proved in analogy with matrix $\hat{C}V$ that, for the linear case, the empirical Ritz eigenvectors, v_1, can be interpreted as the columns of a matrix (Ortega and Messina 2017)

$$O_k = \begin{bmatrix} \tilde{v}_{11} & \tilde{v}_{12} & \cdots & \tilde{v}_{1p} \\ \vdots & \vdots & \ddots & \vdots \\ \tilde{v}_{m1} & \tilde{v}_{m2} & \cdots & \tilde{v}_{mp} \end{bmatrix}. \tag{6.10}$$

Analysis of this expression may provide insight about regions of nonlinearity and observability associated with the fundamental modes of motion.

In cases where the observational data can be associated with a dynamic process measured at different system locations, data points can be visualized as dynamic trajectories whose temporal (spatial) behavior may be highly correlated and exhibit nonlinear dynamics.

As suggested in Equation (6.3), spatiotemporal behavior can be approximated by a summation of block-matrices of the form $\hat{x}_j = a_j(t)\psi_j^T(x)$.

6.5 Dimensionality Reduction for Classification and Cluster Validation

Dimensionality reduction is often used as a preprocessing step in classification, regression, and cluster validation (Lunga et al. 2014). Once a low-dimensional model is obtained, classification and cluster validation techniques can be applied.

6.5.1 Cluster Validation

Clustering can be applied to data visualization, data organization, and exploratory data analysis to name a few applications. In extracting clusters, methods are also needed to measure/validate the performance of cluster algorithms, especially for large-dimensional data bases.

Table 6.1 summarizes some typical cluster quality measures. Excellent accounts and further analytical approaches are given in column two of this table.

6.5.2 Temporal Correlations

Recent studies suggest that the presence of strong temporal correlations may affect the performance of nonlinear dimensionality reduction techniques (Schoeneman et al. 2018). An implicit assumption in some of these techniques is that the input data is weakly correlated.

Data correlation implies that the observation matrix X is not full rank. This also implies that features of the data share information, which may result in spurious information in the data pattern (Ghojogh et al. 2019).

TABLE 6.1

Cluster Quality Measures

Method	References
Davies–Bouldin index	Davies and Bouldin (1979)
Dunn index	Dunn (1974)
Haikidi indexes	Haikidi and Vazirgiannis (2002)
Calisnki–Harbasz criterion	Calinski and Harbasz (1974)
Gap statistics	Albalate and Suendermann (2009)
Silhouette	Rousseeuw (1987)
Root-mean-square standard total deviation, RMD, etc.	Hennig et al. (2016)

Experience from this research shows that linear analysis techniques may overestimate the latent dimension of the process and fail due to the highly correlated nature of dynamic data.

The issue of temporal correlation is especially important in the case of dynamic processes represented by dynamic trajectories, such as in the case of transient simulations. Few attempts have been made to study these issues in the power system literature.

6.5.3 Time-Scale Separation of System Dynamics

Detailed simulation results reveal that projection-based models generate accurate estimates of modal parameters even in the presence of measurement noise and can be used to extract and create reduced representations of large-scale dynamic processes characterized by separation of time scales (Arvizu and Messina 2016; Berry et al. 2013).

In the spirit of spectral methods, a general model of the form

$$\widehat{X} = \begin{bmatrix} \hat{x}_1 & \hat{x}_2 \dots & \hat{x}_d \end{bmatrix} = \underbrace{\sum_{j=0}^{d} a_j(t)\psi_j^T}_{\text{Relevant system behavior}} + \underbrace{\sum_{k=d+1}^{m} a_k(t)\psi_k^T}_{\text{Nonessential coordinates}} \qquad (6.11)$$

is adopted, where d denotes the fast modes of interest and the model (6.11) is estimated based on the frequency range or time scale of interest.

6.6 Markov Dynamic Spatiotemporal Models

6.6.1 First-Order Gauss-Markov Process

Markov state models are a powerful approach to model transient processes and can be used to cluster data. Following Chodera et al. (2007), assume that a process is observed at times $t = 0, \Delta t, 2\Delta t, \dots$ where Δt denotes the observation interval. The sequence of observations can be represented in terms of the state the system visits at each of these discrete times; the sequence of states produced is a realization of a discrete-time.

These models assume that the probability vector at some time $n\Delta t$ can be represented by a model of the form

$$p(n\Delta t) = \begin{bmatrix} P(n\Delta t) \end{bmatrix} p(0), \qquad (6.12)$$

where $P(n\Delta t)$ is the probability matrix of the Markov chain and $p(0)$ is the initial probability vector (Moghadas and Jaberi-Douraki 2019; Swope and Pitera 2004). This implies that the future state depends on the current state and the observational data.

For this process to be described by a Markov chain, it must satisfy the Markov property, whereby the probability of observing the system in any state in the sequence is independent of all but the previous state. Use of this model gives

$$p(n\tau) = \left[P(\tau)\right]^n p(0) \qquad for \ \ 1 < n < m, \tag{6.13}$$

where $\left[P(\tau)\right]^n$ denotes higher powers of the transition matrix.

The following general characteristics of this method can be noted:

1. Matrix P is symmetric and positivity preserving, $(m_{ij} > 0)$.
2. From the properties of a Markov matrix, Matrix P is row-stochastic (i.e., $(m_{ii} = 1, \ \forall i)$, with $0 \le m_{ii} \le 1$).
3. P is positive semidefinite. Further, since P is a Markov matrix, all its eigenvalues are smaller than or equal to 1 in absolute value.
4. The eigenvector spectrum of the transition probability matrix gives information about transitions between subsets and the time scales at which the transitions occur. More precisely, the time scale for a given transition can be calculated as

$$K = -\frac{\tau}{\ln \lambda},$$

where τ is the lag time and λ is the eigenvalue (Noé et al. 2006).

As a result, all eigenvalues λ_j of P are real and the eigenvectors form a basis. This latter feature is exploited in Section 6.6.3 to cluster power system data.

6.6.2 First-Order Markov Differential Equation

Insight into the nature of Markov processes can be obtained from the study of a memoryless master equation. The general form of a first-order Markov differential equation is given by

$$\dot{p}(t) = Kp(t), \tag{6.14}$$

where $p(t)$ is an n-dimensional vector containing the probability to find the system in each of its n states at time t. The matrix $K = \left[k_{ij}\right]$ is a rate matrix, where k_{ij} represents the transition rate constant from state i to state j. As discussed by Noé et al. (2016), the system dynamics can be described by a discrete-time Markov process using the transition matrix $T(\tau)$.

To formalize the model, let p (the vector of transition probabilities) be defined as

$$p(t) = e^{Kt}p(0). \tag{6.15}$$

One way to model the dynamics of trajectories is to rephrase (6.15) in the form

$$p(\tau) = p_o(\Delta t)P(\tau - \Delta t) + \sum_{k=1}^{\infty} p_k(\Delta t)P(\tau - \Delta t), \qquad (6.16)$$

where $p_o = p(0)$ is the probability that the walker remains in its current position after the time step Δt (Wang and Ferguson 2018).

The transition matrix, therefore, provides complementary information to that available from static approaches and can be used for clustering dynamic analysis of memoryless systems.

6.6.3 Markov Clustering Algorithm

The Markov clustering method is a recent addition to clustering dynamic trajectories (Enright et al. 2002). The method is based on the analysis of a transition network, which is obtained by (i) Mapping the dynamic trajectories onto a discrete set of microstates, and (ii) Building a transition network in which the nodes are the microstates and a link is placed between them if two microstates are visited one after the other along the trajectory.

To formalize the adopted model, let $X \in \mathfrak{R}^{m \times N}$ be the matrix of measurements corresponding to a given operating scenario.

The Markov clustering algorithm is summarized in the Markov Clustering Algorithm box.

Markov Clustering Algorithm

Given a trajectory matrix $X \in \mathfrak{R}^{m \times N}$
1. Obtain the undirected graph using the procedures in Chapter 3.
2. Build a transition matrix, $P = \begin{bmatrix} p_{ji} \end{bmatrix} \in \mathfrak{R}^{m \times N}$, in which each element p_{ji} represents the transition probability from node j to node i. Normalize the matrix.
3. Expand by taking the ith power of the matrix

$$(M)^2 = M \times M$$

and normalize each column to one.
4. Inflate each of the columns of matrix M with power coefficient r as

$$(\Gamma_r M)_{pq} = \frac{(M_{pq})^r}{\sum_{i=1}^{k} (M_{pq})^r},$$

where Γ_r is the inflation operator.
5. Repeat steps (3) and (4) until MCL converges to a matrix $M_{MCL}(r)$.
6. Compute the first few right eigenvalues and eigenvectors of the resulting matrix p using Equation (6.4). This results in a set of real eigenvalues and eigenvectors u_i.

As discussed in Cazade et al. (2015), the parameter r determines the granularity of the clustering. This approach can be used to identify state subgroups sharing common, dynamically meaningful characteristics. Experience shows that a relatively low value of r may be enough to capture the dynamics of interest. Several variations of this method have been discussed in the literature.

The next example illustrates the application of this method.

Example 6.1

In this example, the measured data from Section 3.6 is used to illustrate the application of the Markov clustering algorithm. The data set is the same used in Figure 3.11 in Section 3.6.

Using the enumerated procedure, the transition matrix, P, is obtained as

$$P = D^{-1}K \in \mathfrak{R}^{14 \times 14}$$

where D and K are as defined in Section 3.4.

The transition matrix for this example is

$$P = \begin{bmatrix} P_{11} & P_{12} \\ P_{21} & P_{22} \end{bmatrix}$$

with

$$P_{11} = \begin{bmatrix} 0.1273 & 0.0668 & 0.0668 & 0.0668 & 0.0670 & 0.0670 & 0.0670 \\ 0.0668 & 0.1274 & 0.0667 & 0.0667 & 0.0667 & 0.0670 & 0.0670 \\ 0.0668 & 0.0667 & 0.1273 & 0.0667 & 0.0667 & 0.0670 & 0.0669 \\ 0.0667 & 0.0667 & 0.0667 & 0.1274 & 0.0667 & 0.0670 & 0.0669 \\ 0.0668 & 0.0667 & 0.0667 & 0.0667 & 0.1274 & 0.0670 & 0.0674 \\ 0.0670 & 0.0670 & 0.0670 & 0.0670 & 0.0670 & 0.1275 & 0.0670 \\ 0.0670 & 0.0670 & 0.0669 & 0.0669 & 0.0670 & 0.0670 & 0.1275 \end{bmatrix}$$

$$P_{12} = \begin{bmatrix} 0.0674 & 0.0674 & 0.0674 & 0.0673 & 0.0674 & 0.0674 & 0.0673 \\ 0.0674 & 0.0674 & 0.0674 & 0.0673 & 0.0674 & 0.0675 & 0.0674 \\ 0.0674 & 0.0674 & 0.0674 & 0.0673 & 0.0674 & 0.0675 & 0.0674 \\ 0.0674 & 0.0674 & 0.0674 & 0.0673 & 0.0674 & 0.0675 & 0.0674 \\ 0.0674 & 0.0674 & 0.0674 & 0.0673 & 0.0674 & 0.0675 & 0.0674 \\ 0.0673 & 0.0673 & 0.0673 & 0.0672 & 0.0672 & 0.0672 & 0.0672 \\ 0.0673 & 0.0673 & 0.0673 & 0.0672 & 0.0672 & 0.0672 & 0.0672 \end{bmatrix}$$

$$
P_{21} = \begin{bmatrix}
0.0674 & 0.0674 & 0.0674 & 0.0674 & 0.0674 & 0.0673 & 0.0673 \\
0.0674 & 0.0674 & 0.0674 & 0.0674 & 0.0674 & 0.0673 & 0.0673 \\
0.0674 & 0.0674 & 0.0674 & 0.0674 & 0.0674 & 0.0673 & 0.0673 \\
0.0673 & 0.0673 & 0.0673 & 0.0673 & 0.0673 & 0.0672 & 0.0672 \\
0.0674 & 0.0674 & 0.0675 & 0.0674 & 0.0674 & 0.0672 & 0.0672 \\
0.0675 & 0.0675 & 0.0675 & 0.0675 & 0.0675 & 0.0672 & 0.0672 \\
0.0673 & 0.0674 & 0.0674 & 0.0674 & 0.0674 & 0.0672 & 0.0672
\end{bmatrix}
$$

$$
P_{22} = \begin{bmatrix}
0.1274 & 0.0666 & 0.0666 & 0.0668 & 0.0670 & 0.0670 & 0.0671 \\
0.0666 & 0.1273 & 0.0665 & 0.0668 & 0.0671 & 0.0670 & 0.0671 \\
0.0666 & 0.0665 & 0.1273 & 0.0668 & 0.0661 & 0.0670 & 0.0671 \\
0.0668 & 0.0668 & 0.0668 & 0.1275 & 0.0671 & 0.0671 & 0.0672 \\
0.0670 & 0.0671 & 0.0671 & 0.0671 & 0.1273 & 0.0664 & 0.0665 \\
0.0670 & 0.0670 & 0.0670 & 0.0671 & 0.0664 & 0.1272 & 0.0665 \\
0.0671 & 0.0671 & 0.0671 & 0.0672 & 0.0665 & 0.0665 & 0.1273
\end{bmatrix}
$$

Application of the Markov clustering algorithm results in 14 modes; the algorithm takes four steps to converge ($p = 4$). Figure 6.4 shows the clusters extracted using this approach. Comparison of the spatial pattern for the most dominant mode with that obtained using DMs in Figure 6.5 shows the accuracy of the developed models.

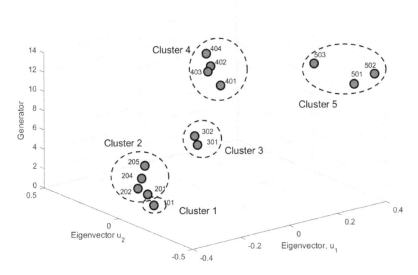

FIGURE 6.4
Schematic illustration of dimensionality reduction. Values unnormalized.

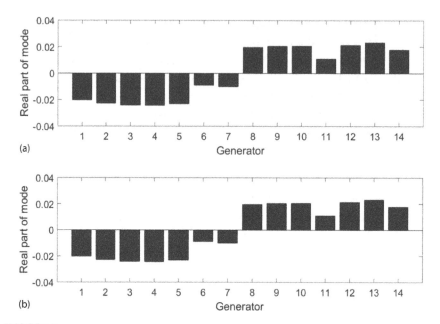

FIGURE 6.5
Schematic illustration of dimensionality reduction. Values unnormalized. (a) Diffusion maps
(b) Markov chain.

6.6.4 Predictive Modeling

All major analytical techniques described above can be used for predictive
modeling. The idea is simple and can be outlined as follows:

1. Given a set of dynamic trajectories, build a low-dimensional representation using a dimensionality reduction technique.
2. Compute spatial and temporal coefficients and build a spatiotemporal representation.
3. Approximate the solution at all future times.

A typical representation model of the output signal $y(t)$ that takes into
account the time-varying and possibly nonlinear behavior of the observed
data is

$$y(t) = m(t) + S(t) + e(t),$$
(6.17)

where $S(t) = \sum_{j=1}^{n} A_j(t)\cos(\omega_j(t))$.

In the above expression, $y(t)$ is the observed output time series, $m(t)$ is a
trend or low frequency component, $S(t)$ is an oscillatory or quasi-oscillatory
component, and $e(t)$ denotes noise. Predictions can be obtained for each

individual oscillatory component $S(t)$, $j = 1, \ldots, n$ and the total model prediction, $\hat{y}(t)$, is obtained by combining the individual predictions.

6.7 Sensor Selection and Placement

Previous studies have suggested that a few diffusion coordinates are often enough to describe the observed global system dynamics. In this section, the sensor selection problem is introduced in the context of dimensionality reduction techniques. This is followed by a discussion of extrapolation.

6.7.1 Sensor Placement

Recent analytical experience by the author using model reduction techniques suggests that the spatial patterns ψ_j (extracted using nonlinear projection methods) can be used to detect good observables or preliminary placement of sensors and impact the determination of relevant signals for sensor placement and state reconstruction. This may be useful to reduce the dimension of the data and/or identify or eliminate redundancy. The latter problem has been recently studied using analytical techniques (Lee 2018).

In Messina (2015) techniques to locate PMUs in order to enhance observability of critical oscillatory modes were introduced. Sensor placement to monitor global system behavior involves the solution of three interrelated problems (Zhang and Bellingham 2008): (a) Determining a limited number of sensors to observe global oscillatory behavior, (b) Selecting the most appropriate type of measurement variables or signal sources so as to have an effective description of the system based on a small number of coarse variables, and (c) Data fusion as in Arvizu and Messina (2016). In addition, it is desirable to avoid redundant measurements to improve the performance of wide-area monitoring systems and extract maximum useful information. A related but separate question is how to reconstruct the observed system behavior from a small number of sensors (field reconstruction) for monitoring and forecasting of wide-area dynamic processes (Usman and Faruque 2019). Recent literature discusses the applicability of these approaches to real-world power system models.

Attention in Section 6.7.2 is centered on the solution of issues (a) and (b).

6.7.2 Mode Reconstruction

Given a set of measurements $X = \{x_i\}_{i=1}^{N}$, a problem of interest is constructing a measurement matrix (a multisensor observing network) $P_{obs} \in \Re^{p \times N}$, or equivalently, determining a subset of sensor locations $x_{obs} = \{x_{k_1} \quad x_{k_2} \cdots \quad x_{k_p}\}$, such that $x_{obs} = P_{obs}X$ with $p \ll m$ that accurately represents the original data set (Zhang and Bellingham 2008; Castillo and Messina 2019). A second objective is to represent the *m*-dimensional measured data by a low-dimensional modal decomposition that captures critical oscillatory behavior of concern; the low-dimensional model can then be used for prognosis, forecasting, and the development of early warning system. Conceptually, this is equivalent to finding good observables that best describe the state of the original system. As a byproduct, the measurement space can be used for estimating system behavior at unobserved system locations.

A simple illustration of this idea is given in Figure 6.6.

In practical applications, matrix P_{obs} is determined by the user based on several criteria such as experience or the application of correlation measures. Note that a related issue was discussed in Chapter 4 in the context of compressive sampling.

The discussion hypothesizes that the selected signals x_{obs} can be determined from the application of nonlinear dimensionality reduction techniques. The proposed approach consists of two main stages: (a) Extracting a low-dimensional representation using any of the dimensionality reduction techniques described in Chapter 3, and (b) Identifying observables from the spatial structure of the low-dimensional embedding.

For reference and comparison, a guided search method is adopted to place sensors (Messina 2015; Alonso et al. 2004). An alternative approach for determining sensor placement has been recently proposed by Castillo and Messina (2019).

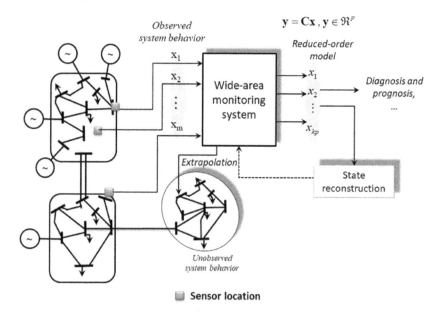

FIGURE 6.6
Generic model for state reconstruction and data interpolation. (Based on Castillo and Messina, 2019).

To optimize the number (and location) of sensors, the error between the approximate field obtained using information from a few sensors (a subspace X_d), $d < M$ and the measured data, X, is calculated. The first problem considered consists of calculating the lowest possible estimation error for the field by using a set of leading modes, assuming that, the first modes retain most of the energy present in the system. In other words, the aim is reducing the dimensionality of data by finding dimension $d < D$ on which the original data can be projected with a small error $X \approx X_d + \varepsilon_d$, where ε_d is the error associated with the approximation.

It can be shown (Castillo and Messina 2019) that the error that minimizes ε_d leads to find the smallest average (mean-squared) distance which measures the difference between the original field X and its projection X_d, namely

$$\varepsilon_{d\_min} = \mathrm{E}\left[\|X - X_d\|^2\right],$$

where ε_{d_min} is the error associated with the $(d - D)$ unimportant or unphysical modes that are discarded.

Example 6.2 will illustrate these ideas.

Example 6.2

As an example, data sets of bus voltage magnitude deviations shown in Figure 6.7 for the Australian test power system were used to identify a limited number of measurements that characterize the overall system behavior. Based on the system response, the data matrix is defined as $X \in \mathfrak{R}^{N \times m} = \begin{bmatrix} V_1 & V_2 \dots & V_m \end{bmatrix}^T$, with $m = 174$ and $N = 3000$.

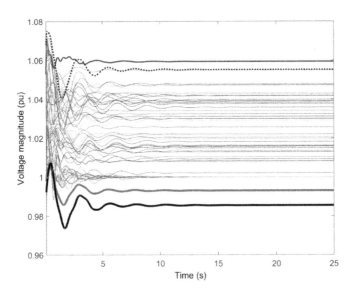

FIGURE 6.7
Bus voltage magnitude deviations following a three-phase short circuit at bus 217.

A two-step approach is utilized to detect good observables to characterize global motion: (a) Introduce a critical contingency that stimulates electromechanical modes in a frequency range of concern $\omega_{min} \leq \omega \leq \omega_{max}$, and (b) Locate sensors to observe the dynamic of interest.

Results involve a three-phase stub fault at bus 217, cleared after 0.05 seconds; this fault is found to excite the two slowest modes in the system. The data set contains 30 seconds of bus voltage magnitude measurements at 59 system locations. Both, load buses and generator terminal buses are selected for analysis.

Table 6.2 lists optimal sensor locations using the approach shown in Messina (2015). In this table, the second column gives the cumulative energy. The technique identifies buses 308, 315, 300 and 509 as the best options to locate sensors.

As discussed in Section 6.4, dimensionality reduction provides a simple approach to estimate good observables for state reconstruction and sensor placement.

Figure 6.8 shows the time evolution of the dominant Koopman modes extracted using the procedure in Chapter 4, while Figure 6.9 shows the voltage-based patterns extracted from the bus voltage measurements.

Figure 6.6 shows the voltage-based shape extracted from the time series in Figure 6.7 using Koopman analysis. For the sake of clarity and completeness, voltage-based shapes are obtained for the three slowest modes in the system. Results suggest that candidate sensor locations are associated with

TABLE 6.2

Sensor Placement Results. Time Window: 1000–3000 s

Candidate Sets	Cumulative Energy
308, 315, 30, 509, 20, 46, 407, 301, 1, 406	99.97%

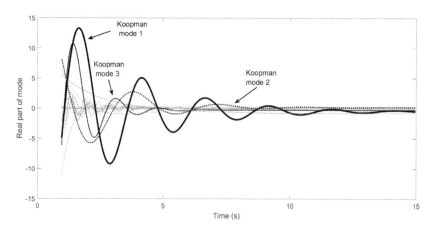

FIGURE 6.8
Dominant Koopman modes.

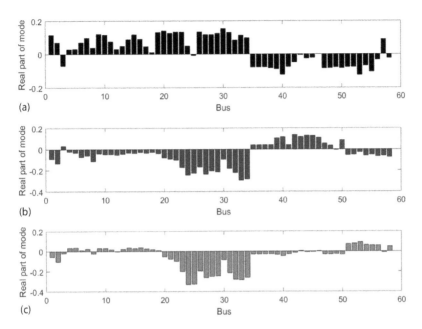

FIGURE 6.9
Voltage-based shapes associated with the Koopman modes in Figure 6.6. (a) Koopman mode 1,
(b) Koopman mode 2, and (c) Koopman mode 3.

the dominant modes excited by the perturbation. This observation is substantially confirmed in the analysis of various test systems and operating conditions.

Based on voltage measurements, results show that dynamic reduction can be useful as an exploratory data mining tool for measured power system data.

6.7.3 Extrapolation

As a motivation for spatial interpolation, let the time evolution of sensors measurements at location i be given by the POD-Galerkin expansion

$$x_i(t) = x(i,t) = \sum_{j=1}^{p} a_j(t)\phi_j(x), i = 1, \ldots, r \tag{6.18}$$

where r denotes the number of selected measurements. Similar approximations can be obtained using other approaches such as DMD, DHR, or Prony analysis to mention a few techniques.

Let $\{s_j : j = 1, \ldots, K\}$ denote the location of measuring locations. An estimate of the time evolution of sensor location j (a site where no measurements are taken) can be obtained as

$$x(j,t) = \sum_{j=1}^{N} u_{ji} \sum_{k=1}^{K} a_k(t)\phi_k(x_i) \tag{6.19}$$

where the u_{ji} are prediction weights for the site i and the unobserved site j (Munoz et al. 2008). The general framework for data extrapolation is illustrated in Figure 6.10.

If necessary, the data can be standardized using the procedures describe in Chapter 3, in which, for system locations where no measurements are available—see Eq. (3.10), the mean values, $\mu(x_i)$, and standard deviations, $\sigma(x_i)$, have to be estimated.

In the simplest case, weights can be obtained from topological conditions; more elaborate approaches that reduce spatial variability can be obtained using formal approaches such as Krigging, correlation analysis, or intelligent analysis techniques.

6.7.4 Data Classification

Once model reduction has been obtained, data classification techniques can be applied to the reduced model. Common classifiers include techniques such support vector machines and linear discriminant analysis techniques. While these methods can be applied directly to the original unreduced data, the large dimensionality of the models may preclude an in-depth analysis of features of interest.

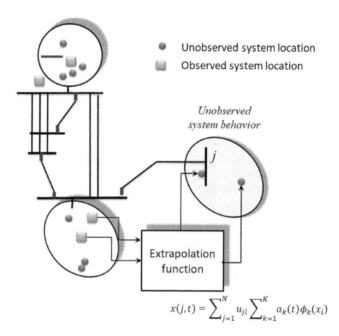

FIGURE 6.10
Extrapolation.

6.8 Open Problems in Nonlinear Dimensionality Reduction

Complex oscillatory processes may contain noise, trends, and other artifacts that can prevent the analysis and extraction of special features of interest, such as localized events. High levels of ambient noise, control actions, switching events, and load variations result in nonstationary behavior and may affect the performance of conventional methods (Strange and Zwiggelaar 2014). Other open problems include missing data and abrupt trend changes.

6.8.1 Effect of Noise and Dropped Data Points

Complex observational data may exhibit nonlinear trends and sudden variations in system behavior (Messina 2015). This nonstationary character of the data makes the analysis and interpretation of system behavior difficult and may lead to biased or incorrect results. Missing data and noise may adversely affect the performance of dimensionality reduction methods (Allen et al. 2013).

There has been little research into the application of smoothing and denoising procedures to power system data. Among these techniques, Wavelet shrinkage and Hilbert–Huang transform analysis have recently emerged as useful signal processing tools for recovering signals from noisy observations. For completeness, these methods are briefly reviewed in the next few paragraphs.

Consider the problem of recovering a function $f(.)$ from noise contaminated observations

$$y_i(t_i) = f(t_i) + \sigma z_i, \qquad i = 1, \ldots, N \tag{6.20}$$

without assuming any particular parametric structure on its form, where y_i is the observed data point, $f(\cdot)$ is the unknown function of interest, t_i are equally spaced points, z_i is standard Gaussian white noise, and σ is a noise level that may, or may not, be known.

In essence, the problem now becomes that of estimating f with small mean-square error. Let $f = f(t_i)_{i=1}^n$ and $\hat{f} = \hat{f}(t_i)_{i=1}^n$ denote the vectors of the true (uncontaminated) signal and estimated sample values, respectively. Minimizing the difference between these two vectors is accomplished by optimizing the mean-squared error, L_2. In other words, the norm risk R is

$$R(\hat{f}, f) = \left\| \hat{f} - f \right\|_{L_2}^2 = \frac{1}{n} \sum_{i=1}^n E(\hat{f} - f)^2 \tag{6.21}$$

subject to the condition that with high probability, $\hat{f}$ is at least as smooth as f.

The steps of the wavelet shrinkage method are given in Messina et al. (2006) and are summarized in the Wavelet Shrinkage Algorithm box:

Wavelet Shrinkage Algorithm

Given noise contaminated observations $y_i(t_i)$, $i = 1, \ldots, n$

1. Compute the one-dimensional wavelet transform of the data, $w = T$, where $w = \begin{bmatrix} w_1 & \cdots & w_n \end{bmatrix}^T$ is the vector of wavelet coefficients, and T is the wavelet transform matrix

$$T = \begin{bmatrix} h_{11} & \cdots & h_{1n} \\ \vdots & \ddots & \vdots \\ h_{n1} & \cdots & h_{nn} \end{bmatrix}.$$

2. Compute an estimate of the standard deviation $\hat{\sigma}$ of the noise in Equation (6.12) from the wavelet coefficients, w.

3. Apply a hard or soft threshold function δ_{λ_k} to the wavelet coefficients w_k

$$\widehat{w}_k = \hat{\sigma} \delta_{\lambda_k}\left(\frac{w_k}{\hat{\sigma}}\right).$$

4. Compute the inverse wavelet transform of the wavelet coefficients

$$f_\lambda = T^{-1}\widehat{w},$$

where the bold font indicates a vector or matrix, thus it is noted that, among other types of smoothing approaches, wavelet shrinkage allows removing noise at all frequencies.

An alternative to filling in missing data can be obtained from Hilbert-Huang analysis. This method decomposes a non-linear time-varying signal into the form

$$x(t) = \sum\nolimits_{i=1}^{n} x_i(t) = \sum\nolimits_{i=1}^{n} A_i(t)\cos\left(\omega_i(t) + \varphi_i(t)\right) \tag{6.22}$$

where $A_i(t)$ is the instantaneous temporal amplitude and $\omega_i(t)$, $\varphi_i(t)$ are the instantaneous frequency and phase, respectively.

Application of the Hilbert-Huang transform results in

$$z(t) = \sum\nolimits_{i=1}^{n} x_i(t) + jH\left[x_i(t)\right] = \sum\nolimits_{i=1}^{n} A_i(t)e^{i\int_o^t \omega_i(t)dt} \tag{6.23}$$

where $H\left[x_i(t)\right]$ is the Hilbert transform of $x_i(t)$.

Through a process called empirical mode decomposition (EMD), the measured signal can be decomposed into nonlinear time-varying components associated with a given frequency or frequency range. Because the different modes of oscillations are separated, filtering can be performed on both time and frequency domains simultaneously. Experience with the analysis of complex measured data shows that general arbitrary signal corrupted with noise can be decomposed into the general form

$$x(t) = \sum_{i=1}^{n} c_i(t) = \underbrace{\sum_{j=1}^{p} c_j(t)}_{\substack{Noise + high \\ frequency \\ components}} + \underbrace{\sum_{k=p+1}^{r} c_k(t)}_{\substack{Physically \\ meaningful \\ components}} + \underbrace{\sum_{l=r+1}^{n} c_l(t)}_{\substack{Trend \ and \\ artificial \\ components}} \tag{6.24}$$

Missing points in measured data are interpreted as very high-frequency components (HFC), which are presented as

$$c_{j\text{HFC}}(t) = Re\left[\sum_{j=1}^{p} \varphi_j e^{(\sigma_j + i\omega_j)t} \right]$$

It follows from Equation (6.18) that the filtered data can be represented as

$$\hat{x}(t) \approx x(t) - \sum_{j=1}^{p} c_j(t) = \sum_{k=p+1}^{r} c_k(t) + \underbrace{\sum_{l=r+1}^{n} c_l(t)}_{} \tag{6.25}$$

The efficiency of the denoising method depends on various factors such as the wavelet basis and the threshold selection method.

6.8.2 Missing Data and Uncertain Value

Missing (dropped) data points are a commonly occurring problem in measured data (Allen et al. 2013, Haworth and Chen, 2012). An example of a measured time series containing missing data recorded using PMUs is shown in Figure 6.11. In fairly simple cases, missing data may be filled in by taking an average of the two preceding samples (Dahal and Brahma 2012). In more complex cases with data with many missing points, however, specialized techniques such as gappy POD (Everson and Sirovich 1995), and Kalman filtering can be used to fill in the missing data and smooth observed behavior (Bui-Than et al. 2004).

Other possibilities are the use of smoothing techniques embedded in the data mining, compressive sampling, data fusion algorithms, the application of time decomposition methods such as the Hilbert–Huang transform, and DHR. Figure 6.11 shows an example of a measured signal exhibiting a significant number of missing points. Figure 6.12 compares the extracted signals obtained using HHT, Equation (6.18), and DHR. Results match well with the original data with multiple missing points.

Results show that these techniques can be efficiently used to fill in missing data. Similar results are obtained using other techniques and are not reported here.

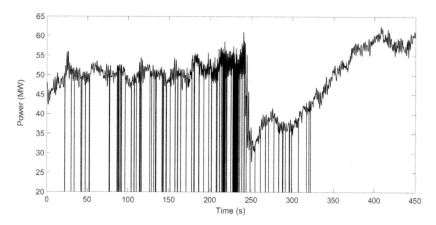

FIGURE 6.11
Measured signal containing missing data points.

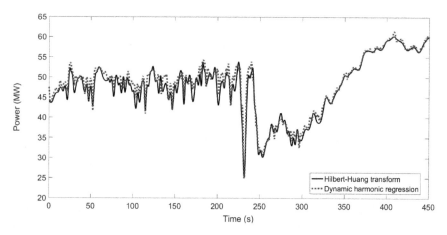

FIGURE 6.12
Reconstructed data using HHT and DHR.

References

Albalate, A., Suendermann, D., A Combination approach to cluster validation based on statistical quantiles, *2009 International Joint Conferences on Bioinformatics, Systems Biology and Intelligent Computing*, Shanghai, China, August 2009.

Allen, A., Santoso, S., Muljadi, E., Algorithm for Screening Phasor Measurement Unit Data for Power System Events and Categories and Common Characteristics for Events Seen in Phasor Measurement Unit Relative Phase-Angle Differences and Frequency Signals, Technical Report, NREL/TP-5500-58611, August 2013.

Alonso, A. A., Frouzakis, C. E., Kevrekidis, I. G., Optimal sensor placement for state reconstruction of distributed process systems, *Process Systems Engineering*, 50(7), 1438–1452, 2004.

Arvizu, C. M. C., Messina, A. R., Dimensionality reduction in transient simulations: A diffusion maps approach, *IEEE Transactions on Power Systems*, 31(5), 2379–2389, 2016.

Barocio, E., Pal, B., Thornhill, N. F., Messina A., A dynamic mode decomposition framework for global power system oscillation analysis, *IEEE Transactions on Power Systems*, 30(6), 2902–2912, 2015.

Berry, T., Cressman, R., Ferencek, G. Z., Sauer, T., Time-scale separation from diffusion-mapped delay coordinates, *SIAM Journal on Applied Dynamical Systems*, 12(2), 618–649, 2013.

Bui-Than, T., Damodaran, M., Willcox, K., Aerodynamic data reconstruction and inverse design using proper orthogonal decomposition, *AIAA Journal*, 8(42),1505–1516, 2004.

Calinski, T., Harbasz, J., A dendrite method for cluster analysis, *Communications in Statistics*, 3(1), 1–17, 1974.

Castillo, A., Messina, A. R., Data-driven sensor placement for state reconstruction via POD analysis, *IET Generation, Transmission & Distribution*, 14(4), 656–664, 2020.

Cazade, P. A., Zheng, W., Prada-Garcia, D., Berezovska, B., Rao, F., Clementi, C., Meuwly, M., A comparative analysis of clustering algorithms: O_2 migration in truncated hemoglobin I from transition networks, *The Journal of Chemical Physics*, 142, 025103, 2015.

Chan, S. M., Modal controllability and observability of power system models, *Electric Power & Energy Systems*, 6(2), 83–88, 1984.

Chen, Y. Q., Xie, L., Kumar, P. R., Dimensionality reduction and early event detection using online synchrophasor data, *2013 IEEE Power & Energy Society General Meeting*, Vancouver, CA, July 2013.

Chodera, J. D., Singhal, N., Pande, V. S., Dill, K. A., Swope, W.C., Automatic discovery of metastable states for the construction of Markov models of macromolecular conformational dynamics, *The Journal of Chemical Physics*, 126(155101), 1–17, 2007.

Dahal, O. P., Brahma, S. M., Preliminary work to classify the disturbance events recorded by Phasor measurements units, *2012 IEEE Power and Energy Society General Meeting*, San Diego, 2012.

Davies, D. L., Bouldin, D. W., Cluster separation measure, *IEEE Transactions on Pattern Analysis and Machine Intelligence*, 1(2), 95–104, 1979.

Dunn, J. C., Well separated clusters and optimal fuzzy partitions, *Journal of Cybernetics*, 4(1), 95–104, 1974.

Dzemyda, G., Kurasova, O., Zilinskas, J., *Multidimensional Data Visualization – Methods and Applications*, Springer, New York, 2013.

Enright, E. J., Dongen, S. V., Ouzounis, C. A., An efficient algorithm for large-scale protein families, *Nucleic Acids Research*, 30(7), 1575–1584, 2002.

Erban, R, Frewen T. A., Wang, X., Elston, T. C., Coifman, R., Nadler, B., Kevrekidis, I. G., Variable-free exploration of stochastic models: A gene regulatory network example, *The Journal of Chemical Physics*, 126(5), 155103, 2007.

Everson, R., Sirovich, L., Karhunen-Loeve procedure for gappy data, *Journal of the Optical Society of America A*, 12(8), 1657–1664, 1995.

Fodor, I. K., A survey of dimension reduction techniques. Technical Report UCRL-ID-148494, Lawrence Livermore National Laboratory, http://computation.llnl.gov/casc/sapphire/pubs/148494.pdf, 2002.

Ghojogh, B., Samad, M. N., Mashhadi, S. A., Kapoor, T., Ali, W. F., Karray, F., Crowley, M., Feature selection and feature extraction in pattern analysis: A literature review, ArXiv:1905.02845, 2019.

Haikidi, M., Vazirgiannis, M., Cluster validity assessment using multirepresentatives, *2nd Hellenic Conference on AI, SETN-2002*, Thessaloniki, Greece, pp. 237–248, April 2002.

Haworth, J., Cheng, T., Non-parametric regression for space-time forecasting under missing data, *Computers, Environments and Urban Systems*, 36, 538–550, 2012.

Hennig, C., Meila, M., Murthag, F., Rocci, R., *Handbook of Cluster Analysis*, Chapman & Hall/CRC, Boca Raton, 2016.

Lee, B., Multigrid for model reduction of power grid networks, *Numerical Linear Algebra Applications*, 25(3), 1–27, 2018.

Lee, J., Verleysen, M., *Nonlinear Dimensionality Reduction, Information Science and Statistics*, Springer, New York, 2007.

Li, Y., Li, G., Wang, Z., Han, Z., Bai, X., A multifeatured fusion approach for power system transient stability assessment using PMU data, *Mathematical Problems in Engineering*, 2015, 786396, 2015.

Long, A. W., Phillips, C. L., Jankowksi, E., Ferguson A. L., Nonlinear machine learning and design of reconfigurable digital colloids, *Soft Matter*, 12(34), 7119–7135, 2016.

Lu, H., Plataniotis, K. N., Venetsanopoulos, A. N., *Multilinear Subspace Learning: Dimensionality Reduction of Multidimensional Data*, Machine Learning and Pattern Recognition Series, Chapman & Hall/CRC, Boca Raton, 2014.

Lunga, D., Prasad, S., Crawford, M. M., Ersoy, O., Manifold-learning based feature extraction for classification of hyperspectral data, *IEEE Signal Processing Magazine*, 31(1), 56–66, 2014.

Mendoza-Schcrock, O. L., Patrick, J., Arnold, G., Ferrara (Eds.), Manifold learning and the Caesar database, in *Evolutionary and Bio-inspired Computation: Theory and Applications III Proceedings of Spie*, 8402, 2012.

Messina, A. R., *Wide-Area Monitoring of Interconnected Power Systems, Power and Energy Series*, 77, Institution of Engineering and Technology, Stevenage, UK, 2015.

Messina, A. R., Vittal, V., Ruiz-Vega, D., Enríquez-Harper, G., Interpretation and visualization of wide-area PMU measurements using Hilbert analysis, *IEEE Transactions on Power Systems*, 21(4), 1763–1771, 2006.

Moghadas, S. M., Jaberi-Douraki, M., *Mathematical Modelling*, John Wiley & Sons, Inc., Hoboken, NJ, 2019.

Munoz, M., Lesser, V. M., Ramsey, F. L., Design-based empirical orthogonal function model for environmental monitoring data analysis, *Environmetrics*, 19, 805–817, 2008.

Noé, F., Banisch, R., Clementi, C., Commute maps: Separating slowly mixing molecular configurations for kinetic modeling, *Journal of Chemical Theory and Computation*, 12, 5620–5630, 2016.

Ortega, M. A. H., Messina, A. R., An observability-based approach to extract spatio-temporal patterns from power system Koopman mode analysis, *Electric Power Components and Systems*, 45(4), 355–365, 2017.

Ramos, J., Kutz, J. N., Dynamic mode decomposition and sparse measurements for characterization of power system disturbances, arXiv:1906:03544, 2019.

Rousseeuw, P. J., Silhouettes: A graphical aid to the interpretation and validation of cluster analysis. *Journal of Computational and Applied Mathematics*, 20, 53–65, 1987.

Schoeneman, F., Chandola, V., Napp, N., Wodo, O., Zola, J., Entropy-Isomap: Manifold learning for high-dimensional dynamic processes, *2018 IEEE International Conference on Big Data*, Seattle, USA, December 2018.

Singer, A., Detecting intrinsic slow variables in stochastic dynamical systems by anisotropic diffusion maps, *PNAS*, 106, 16090–16095, 2009.

Strange, H., Zwiggelaar, R., *Open Problems in Spectral Dimensionality Reduction Methods*, Springer Briefs in Computer Science, Springer Cham, New York, Heidelberg, 2014.

Swope, W. C., Pitera, J. W., Describing protein folding kinetics by molecular dynamics simulations, Part 1, theory, *Journal of Physical Chemistry B*, 108, 6571–6581, 2004.

Tenenbaum, J. B., de Silva V., Langford J. C., A global geometric framework for non-linear dimensionality reduction. *Science*, 290(5500), 2319–2323, 2000.

Usman, M. U., Faruque, M. O., Application of synchrophasor technologies in power systems, *Journal of Modern Power Systems and Clean Energy*, 7(2), 211–226, 2019.

Van der Maaten, L., Postma, E., Van den Herik, J., Dimensionality reduction: A comparative review, Tilburg Centre for Creative Computing, TiCCTR2009–005, October 2009.

Van der Matten, L., Postma, E., Jaap Van Den Herik, H., A taxonomy of dimensionality reduction techniques, *Journal of Machine Learning Research*, 10(1), 1–22, 2006.

Wang, J., Ferguson, A. L., A study of the morphology dynamics and folding pathways of ring polymer with supramolecular topological constraints using molecular simulations and nonlinear manifold learning, *Macromolecules*, 51, 598–616, 2018.

Yan, J., Chang, B., Liu, N., Yan, S., Cheng, Q., Fan, W., Yang, Q., Xi, W., Chen, C., Effective and efficient dimensionality reduction for large-scale and streaming data preprocessing, *IEEE Transactions on Knowledge and Data Engineering*, 18(3), 320–333, 2006.

Zhang, Y., Bellingham, J. G., An efficient method for selecting ocean observing locations for capturing the leading modes and reconstructing the full field, *Journal of Geophysical Research*, 113(C04005), 1–24, 2008.

7

Forecasting Decision Support Systems

7.1 Introduction

Most power control centers have operationally implemented advanced wide-area measurement systems and data processing strategies to monitor power system health. These strategies include the implementation of situational awareness systems including disturbance alert, event location triangulation and oscillation detection, early warning systems, data archiving, and other advanced features (Begovic et al. 2005; Huang et al. 2010).

Forecasting plays an important role in prognosis and decision-making actions by operators. Accurate predictions of system behavior can be used to provide early warning of impending events and improve system security. This information is necessary for properly responding to system threats in order to mitigate the impact of these threats on system operation.

Accurate forecasts depend on how well the initial state or condition can be characterized and rely on the accuracy or fidelity of the forecasting model. Important features of prognostic and decision-making systems allow automated decision support systems to help operators make faster and more effective decisions to detect and respond to recurring conditions in the transportation network.

Forecasting models can be classified by categories that are generally divided into physical models, statistical models (machine learning models), and hybrid physical statistical models.

This chapter reviews the state of the art in forecasting decision and support systems. Emphasis is placed on spatiotemporal prediction and forecasting and its role in decision support. The roles and integration of prognostic algorithms, required to predict future system health, are also discussed.

7.2 Background: Early Warning and Decision Support Systems

Timely detection and accurate identification of temporal variations in the dynamic pattern of system oscillations lies at the heart of developing novel algorithms for instability or threat detection in stressed power systems.

There have been an extensive number of research efforts focused around developing early warning and decision support systems. Figure 7.1 shows a conceptual illustration of a prognostics and decision support system embedded in a WAMS. The components of a decision support system may include:

1. A data fusion system designed to reduce system dimensionality. Building on the data provided by the wide-area monitoring system, the data fusion stage aims at determining a critical number of observables or measurements that best describe system dynamics.

2. A state monitoring stage designed to reconstruct system behavior from a few selected signals. As such, the system must have prognostic as well as diagnostic capability.

3. A prognostics/forecasting stage. For power system monitoring applications, the objective is the early detection of abnormal events and the detection and extent of the propagating phenomena.

4. A decision support system for monitoring, reporting, and forecasting power system health.

Desirable features of an advanced early warning and decision support systems are discussed in North American Electric Reliability Corporation (2010), Sohn et al. (2001) and include:

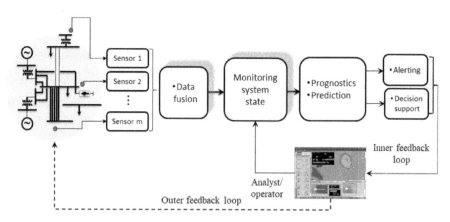

FIGURE 7.1
Envisioned prognostics and decision support system. (Based on Callan, R. et al., An integrated approach to the development of an intelligent prognostic health management system, 2006 *IEEE Aerospace Conference*, Big Sky, MT, March 2006.)

- Geospatial views of the grid (observation),
- Holistic views combining data from multiple diverse sources,
- Monitoring and detection of impending faults,
- Advanced, fast, alert systems to provide the operator with early warning of impending threats,
- Root-cause analysis to quickly identify sources of problems,
- Diagnostic tools that recommend corrective actions, and
- Forecasting techniques to predict imminent system conditions and trigger emergency control actions.

Key to these systems are techniques for data manipulation, condition monitoring, health assessment, prognostics, and automatic decision reasoning (Callan et al. 2006).

Decision support systems have evolved from simple monitoring systems to sophisticated predictive solutions that use state-of-the-art machine learning and simulation technologies to support decision making. The decision support system monitors power system health, detects abnormal events, and suggest diagnostics actions to the operator.

An advanced monitoring system takes the output of a WAMS, performs fault detection, fault diagnosis, and fault prognosis using learning techniques. In addition to fault detection, the data prognosis module also includes support for deciding what actions to take in response to a fault and triggering alarms. These actions can include reconfiguration, alarm setting, and early warning systems; thus the module become an integral part of a health management system. Prognostics is usually recognized as the most difficult part of a health management system (Schwabacher and Goebel 2007) and is likely to become a subject of increasing interest in the future smart grid (Borlase 2012).

By integrating advanced forecasting techniques with prognosis analysis, techniques to provide automatic early detection of system deterioration and the development of control actions can be developed.

Two main loops can be considered in a real-time decision support architecture for monitoring power system behavior:

1. *An inner feedback loop*: The inner feedback loop monitors system response to control actions and includes data fusion, monitoring, and prognostic stages. Forecasting yields a prediction about events and can incorporate techniques to detect the location and severity of the events.

2. *An outer feedback loop*: This loop aims at performing more selective monitoring and may incorporate facilities to improve models and assess the effect of mitigation/control measures. As noted in a recent report published by the National Academies of Sciences,

Engineering, and Medicine (https://www.nap.edu/read/24962/chapter/3#10), the tasks in the outer feedback loop are expected to operate on longer time scales and may include operator intervention.

Typically, the decision stage requires a forecast/prediction over a number of feature periods (i.e., minutes or hours) and may incorporate data fusion and monitoring states. The length of the planning horizon depends upon the nature of the phenomena being studied and the method of analysis being applied. These considerations also determine the architecture needed for decision support.

7.3 Data-Driven Prognostics

An analytical framework that illustrates the use of system forecasts and prognosis to aid in decision support is illustrated in Figure 7.2. The inputs to this model include: relevant system parameters, forecasts of key system variables, real-time data, and historical data. Examples include weather data, same day data, etc. (Mohan and Kumar, 2018).

The methods used in forecasting measured data fall into three major categories: (a) time series models, (b) machine learning methods, (c) deep learning methods, and (d) hybrid techniques.

As discussed in Poslad et al. (2015), critical aspects of the early warning system include:

- Time-critical sensor data exchange, and
- Scalability, (i.e., the need for scalable time-sensitive data exchange and processing).

Examples of techniques used for data-driven prognostics include Kalman filters, wavelets, dynamic mode decomposition, dynamic harmonic regression, particles, and regression. Table 7.1 summarizes some approaches to

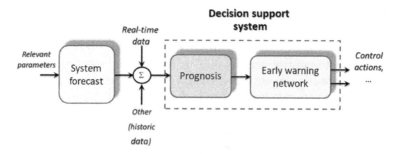

FIGURE 7.2
Pipeline for early warning and decision support system.

TABLE 7.1

Overview of Data-Driven Prognostic Approaches for Health Management

Approach	Categories
Machine learning approaches	Artificial neural networks
	Support vector machines (SVM)
	Decision trees
	Bayesian models
Statistical approaches	Markov models
	ARMA and related techniques
	Proportional hazard models
Time-domain methods and conventional numerical algorithms	Dynamic harmonic regression
	Kalman filtering
	Koopman mode analysis
	Kalman filters
	Wavelets
Hybrid methods	Linear regression, …

Source: Xia, T. et al., *Reliability Engineering and System Safety*, 178, 255–268, 2018.

data-driven prognostics (Xia et al. 2018) in the context of health management for manufacturing systems. Schwabacher and Goebel (2007) provide an overview of data-driven prognostics in the context of integrated systems health management.

Sections 7.4–7.8 examine the utility of forecasting and decision support systems from the perspective of power system monitoring. Emphasis is placed on time domain methods.

7.4 Space-Time Forecasting and Prediction

Forecasting is a key element of decision making (Montgomery et al. 1990). *Time series forecasting* is broadly defined as the process of predicting the future value of time series data based on past output measurements and other inputs (Kotu and Deshpande 2019).

Formally, given observations $y(t) = \{y(1),\dots,y(n)\}$ of the output of a system, forecasting is the prediction of the outputs $y(t) = y_{t+h} = \{y(n+1),\dots,y(n+h)\}$ based on all the data up to time t. Following the notation of Hyndman and Khandakar (2008), the estimated value of the time series y at time $t+h$ is denoted as $\hat{y}_{t+h|t}$.

There are several approaches to time series forecasting that could be implemented in an online prediction system. These methods can be classified in a number of ways which include forecasting based on: time series decomposition, smoothing based techniques, regression based techniques, and machine learning-based techniques (Kotu and Deshpande 2019). Other classifications are described in (Mohan and Kumar 2018).

The succeeding sections briefly examine the application of these techniques to forecast time series behavior. This chapter concentrates on the analysis of measured power system data.

7.4.1 Decomposition-Based Methods

Decomposition-based methods split a time series, $y(t)$, into a combination of fundamental components, typically a trend or slowly varying component, a cyclical or quasi-cyclical (oscillatory) component, and a noise component.

In the analysis that follows, it will be assumed that measured data can be adequately represented by a general model of the form

$$y_t = y(t) = \underbrace{T_t}_{\substack{\text{Slowly} \\ \text{varying} \\ \text{trend}}} + \underbrace{S_t}_{\substack{\text{cyclical} \\ \text{component}}} + \underbrace{e_t}_{\text{noise}} \tag{7.1}$$

where T_t is the temporal trend, S_t denotes the cyclical (oscillatory) or quasi-cyclical component, and e_t represents noise (Young 2011). Examples of such representations include wavelets, Hilbert-Huang analysis, and DHR.

Once the time series is decomposed into its fundamental components, the future value for each component can be calculated and then combined back together for forecasted future time series values. As a by-product, the noise component can be monitored to detect outliers. For example, the periodic components may represent harmonic or modal components as discussed in Zavala and Messina (2014) or dynamic mode decomposition as in Mohan and Kumar (2018). These methods are well developed and are routinely applied to estimate modal behavior.

7.4.2 Exponential Smoothing Methods

Because of their relationship with other multivariate forecasting methods, these techniques are briefly reviewed below. These methods obtain forecasts of a time series based as the weighted moving average of all past observations, where the weights decrease exponentially. This family of methods can be written in the general form

$$\hat{y}_{t+1|t} = l_t$$
$$l_t = \alpha y_t + \alpha(1-\alpha)y_{t-1} + \alpha(1-\alpha)^2 y_{t-2} + \ldots \tag{7.2}$$

where $\hat{y}_{t+1|t}$ is the estimated value of the time series, l_t is the series level at time t, and α is the smoothing parameter between 0 and 1 (Kotu and Deshpande 2019). Variants of this model are discussed in Hyndman et al. (2008). Each term combines error, trend, and seasonal components.

In a broad taxonomy, approaches to exponential smoothing can be divided into two main groups: exponential smoothing with trend and exponential smoothing with periodic components. A more detailed explanation of these models and their relationship with state-space representations is given in Hyndman et al. (2002).

7.4.3 State-Space Models

All exponential smoothing methods can be expressed in their innovations state space forms. Following the general approach of Ljung (1999), an n- dimensional state process $y(t)$ can be represented in state space form as

$$\dot{x}(t) = Ax(t) + Bu(t) + Ke(t)$$
$$y(t) = Cx(t) + Du(t) + e(t) \tag{7.3}$$

where $x(t)$ is an unobserved n-dimensional state process, $e(t)$ is the disturbance, and A, B, C, B and K are fixed-coefficient state-space matrices to be estimated. Discrete versions of this model are examined in the following.

A brief outline of this algorithm is given in boxed form.

Forecasting Time Series Algorithm Using State-Space Models

1. Given a model of the form (7.1). Generate a 1-step ahead predictor model for the identified model of the form

$$\hat{x}(t+1) = (A - KC)\hat{x}(t) + Ky(t)$$
$$\hat{y}(t) = C\hat{x}(t) \tag{7.4}$$

with initial conditions $\hat{x}(0) = x_o, w$, where $y(t)$ is the measured output and $\hat{y}(t)$ is the predicted value. The measured output is available until time step N and is used as an input in the predictor model.

2. Compute $\hat{x}(N+1)$, the value of the states at the time instant $t = N + 1t$, which is the time instant following the last available data sample using the measured observations as inputs, as

$$\hat{x}(1) = (A - KC)x_o + Ky(o)$$
$$\hat{x}(2) = (A - KC)\hat{x}(1) + Ky(1)$$
$$\vdots$$
$$\hat{x}(N+1) = (A - KC)\hat{x}(N) + Ky(N) \tag{7.5}$$

3. Simulate the response of the identified model for H steps using $\hat{x}(N+1)$ as initial conditions, where H is the prediction horizon. This response is the forecasted response of the model.

Note: This model requires specification of several parameters, such as the initial conditions and the prediction horizon.

Once a forecast of system behavior is obtained, various techniques can be applied to detect events and identify which events may be critical.

Example 7.1

As an illustration, measured data from a real event in Figures 7.3 and 7.4 is considered; the time series shows three anomalous events at $t = 125, 180$, and 205 seconds associated with a failed system interconnection (Martinez and Messina 2011). The goal is to forecast the K-step ahead output using the first 350 seconds of the measured time series.

For ease of illustration, it is assumed that the observed process follows an auto-regressive (AR) model. The adopted approach involves two main steps:

1. Build an auto-regressive model of the form $A(q)y(t) = e(t)$, where $A(q)$ is a polynomial in q using the approach in Marple (1987), and

2. Given this model, the k-step ahead forecast was obtained using the state-space based forecasting time series algorithm.

As shown in Figure 7.5, the model is found to provide a good fit to the measured time series even when changes in the system trend are present. The results improve for better behaved signals, such as those associated with slowly damped inter-area oscillations.

A drawback of some existing approaches to system forecasting is that calculations are performed offline, once all the data (or the data in a given time window) have been acquired. Alternatively, a sliding-window approach can be used to approximate system behavior. Discussion of this issue is deferred until Chapter 9.

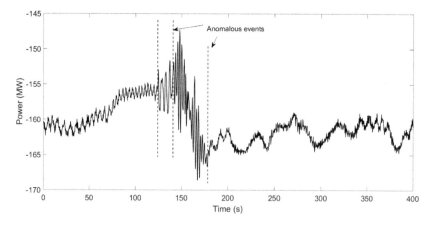

FIGURE 7.3
Time history of measured data.

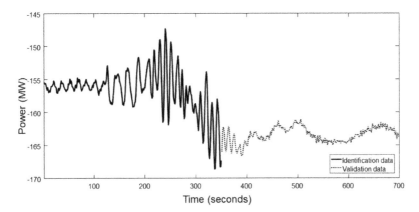

FIGURE 7.4
Definition of identification and validation data.

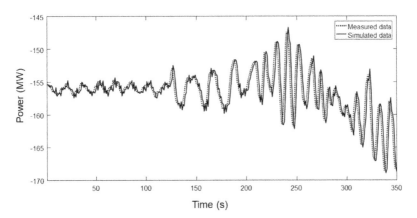

FIGURE 7.5
Out-of-samples prediction. Four steps ahead prediction.

7.5 Kalman Filtering Approach to System Forecasting

Estimation of the time evolution of the state in (7.3) relies initially on application of the Kalman filter. In this section, a Kalman filtering approach to power system monitoring and prediction that combines some of the ideas in the system literature is reviewed. Smoothing methods are also considered and connections to other sequential methods are also highlighted.

7.5.1 The Kalman Filter

The Kalman Filter (KF) recursively estimates the state of a system, x_t, at each time t (once new observations are available) using a state-space model and measurements z_t, where x and y are governed by the linear stochastic difference equations of the form shown in Kalman (1960) and Simo (2014):

$$x_t = Fx_{t-1} + w_t$$
$$y_t = H_t x_t + \xi_t \tag{7.6}$$

In Equation (7.6), x_t is the n-dimensional *state* vector, y_t is the observed data, and ξ_t, w_t represent the measurement noise and zero mean white *process* noises, respectively, with covariances

$$R_t = E\left[w_t w_t^T\right]$$
$$Q_t = E\left[v_t v_t^T\right]$$

where w and ξ are assumed to be Gaussian random vectors with zero mean and covariance matrices $Q \geq 0$ and $R > 0$, respectively.

The discrete Kalman filter equations can be outlined as follows:

a. *Forecasting*

$$\hat{y}_{t|t-1} = H^T \hat{x}_{t|t-1}$$
$$\hat{v}_{t|t-1} = H^T P_{t|t-1} H \tag{7.7}$$

b. *Updating or state filtering*

$$\hat{x}_{t|t} = \hat{x}_{t|t-1} + P_{t|t-1} H \hat{v}_{t|t}^{-1} \left(y_t - \hat{y}_{t|t-1}\right)$$
$$\hat{P}_{t|t} = \hat{P}_{t|t-1} - F\hat{P}_{t|t} H \hat{v}_{t|t-1}^{-1} \hat{P}_{t|t-1} \tag{7.8}$$

c. *Measurement residual*

$$v_t = y_t - \hat{y}_{t|t-1} \tag{7.9}$$

d. *State prediction*

$$\hat{x}_{t+1|t} = F\hat{x}_{t|t}$$
$$\hat{P}_{t+1|t} = F\hat{P}_{t|t} F^T + W \tag{7.10}$$

where $\hat{P}_{t|t}$ is an error covariance matrix of $\hat{x}_{t|t}$ with initial conditions $\hat{x}_o$ and P_o.

Given the state-space model (7.6), noise statistics Q_k and R_k, and initial values of the state vector and its variance-covariance matrix, the Kalman filter can be estimated using Equations (7.7)–(7.10) (Ji and Herring 2013):

In Equations (7.7) through (7.10), Λ_t is the one-step-ahead prediction or innovation, K_t is the Kalman gain at t_{k+1}, $P_{t|t-1}$ is an error covariance matrix of $\hat{x}_{t|t-1}$, and T denotes matrix transpose. In Equation (7.8), the state vector is predicted according to a given state-space model.

The Kalman filter model estimates the parameters of the system defined in Equation (7.6), based on current and past observations by first running a one-step-ahead forecast followed by a correction step. When new data become available, Equation (7.7) provides an estimate of the state vector by combining the model prediction and the data, weighted by the Kalman gain that minimizes the mean square error.

Based on Kalman analysis, kernel regression methods can be applied to the smooth series to determined spatial patterns. Two extensions to this approach are of particular interest to power system oscillation monitoring: forecasting and time-varying trend identification. These extensions are discussed below in Example 7.2 and later in Section 7.6.

Example 7.2

To further illustrate the potential usefulness of these methods, forecasting techniques are applied to active wind power. Figure 7.6 shows actual wind turbine power output measured at the common point of a large wind farm. The measured signal exhibits negative generation ramps.

Kalman filter estimates in Figure 7.7a are found to provide an accurate representation of system behavior. Note that the error estimates increase at the point of strong changes or discontinuities as shown in Figure 7.7b as expected from physical considerations.

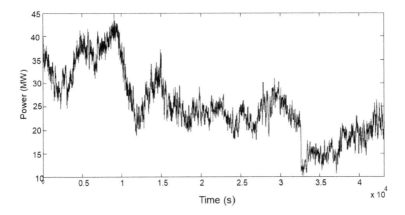

FIGURE 7.6
Measured wind turbine power output.

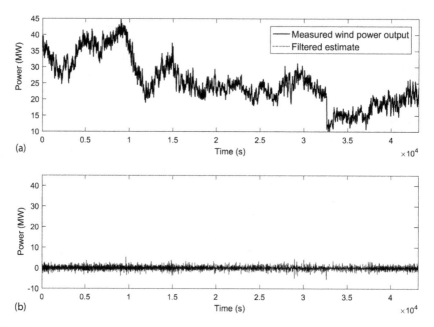

FIGURE 7.7
Kalman estimate of measured wind turbine power output. (a) Kalman filter estimate, (b) estimate error.

7.5.2 Forecasting

As suggested in Section 7.5.1, the Kalman filter can also provide multiple (N) step ahead forecasts. From (7.8), the model prediction is given by

$$\hat{x}_{j_{t+N}} = F_{t+N|t}\hat{x}_t \tag{7.11}$$

where, $F_{t+N|t}$ is the *N-step* ahead transition matrix. Consequently, the *N*-step ahead forecast equation is

$$\hat{x}_{j_{t+N}} = F_{j_{t+N}}\hat{x}_{j_t} \tag{7.12}$$

for $t = 1,\dots,N$, where F_j is the *N*-step ahead transition matrix.

    This forecast equation is kept separate from the recursive Kalman filter algorithm. Applications of this approach to forecast short-term wind output power are discussed in Zavala and Messina (2014).

7.5.3 Time-Varying Trend Estimation

An interesting feature of Kalman analysis is its ability to extract time-varying means. As noted in Chapters 4 and 6, a measured signal can be separated into a smooth time-varying mean and noise as

$$y_t = m_t + e_t, \qquad t = 1, \ldots, N \qquad (7.13)$$

where m_t is the unknown smooth trend and t represents noise, which is assumed to have zero mean.

Assume further, that the trend m_{t+1} at time $t+1$ can be predicted from a previous value, using a simple linear regression

$$m_{t+1} = m_t(t) + Aa_t$$

where $A(t)$ is a lower-order polynomial regression model and it is assumed that $m_t(0)$ is known. In practice a simple three-point estimate (a quadratic filter) can be obtained from

$$m_{t|t-1} = 3(m_{t-1} - m_{t-2}) + m_{t-3}$$

where $m_{t|t-1}$ is the predicted trend value at time t given data through time $t-1$ (Martinelli and Rhoads 2010). In the context of the Kalman filter, the value $m_{t|t-1}$ represents filtered values.

Under this assumption, a final trend estimate can be obtained from

$$m_{t|t} = m_{t|t-1} + K_t\left(y_t - m_{t|t-1}\right) \qquad (7.14)$$

where K_t is the Kalman gain.

7.6 Dynamic Harmonic Regression

Among the structural time series models, dynamic harmonic regression (DHR) has recently attracted attention (Young et al. 1999; Zavala and Messina 2014). In this approach, the time series model illustrated by Equation (7.1) is assumed to consist of three components: (a) A periodic component associated with system oscillatory behavior, (b) A quasi-cyclical component related to the signal trend, and (c) An irregular component that captures non-systematic components. The level of the series and the slope of the trend are assumed to be stochastic. This argument extends previous formulations given in Zavala and Messina (2014) and focuses on the predictive capability of the method.

A general unobserved time components model can be represented by the following equation

$$y_t = T_t + S_t + C_t + e_t = \left(\sum_{j=0}^{R} s_t^{p_j}\right) + e_t$$

where y_t is the observed time series, t denotes time, T_t, S_t, C_t represent the trend, quasi-cyclical (oscillatory) components, and e_t is an irregular component normally distributed Gaussian sequence with zero mean value.

The time-varying components are given by

$$y_t^{p_j} = \sum_{j=1}^{R} C_{j_t} \cos(\omega_j t + \varphi_j) = \left(\sum_{j=1}^{R} a_{j_t} \cos(\omega_j t) + b_{j_t} \sin(\omega_j t) \right) \qquad (7.15)$$

where $C_{j_t} = C_j \cos(\omega_j t + \varphi_j)$, and the $\omega_j = 2\pi f_j, j = 1, \ldots, n$ are the fundamental and harmonic components of the sinusoidal components associated with the jth DHR component. The parameter R determines the number of harmonic regressions that are allowed in the model.

In this representation, a_{j_t} and b_{j_t} are assumed to be stochastic, time-varying parameters (TVP's) that follow a Generalized Random Walk (GRW) process, and e_t is used to represent noise in the observed time series.

The components for $j = 1, 2, \ldots, R$ are named as the different DHR oscillatory components; when $j = 0$, the only remaining term in the summation is s_o which represents a slowly varying component or trend.

Parametric models of the form illustrated by Equation (7.15) are of interest when the trend and cyclical components tend to be periodic, for instance, resulting from forcing.

7.6.1 State Space Representation

Using the state space framework, the DHR model becomes for one cycle ($R = 1$),

$$y_{j_t} = \Lambda_t \eta_{j_t} + e_{j_t} = \begin{bmatrix} \cos\omega_{j_t} & \sin\omega_{j_t} \end{bmatrix} \begin{bmatrix} a_{1j_t} \\ b_{1j_t} \end{bmatrix} \qquad (7.16)$$

Extension to the multiple cyclic process is immediate and is discussed in Chow et al. (2009).

Expanding Equation (7.16) results in the state-space representation

$$\begin{aligned} x_t &= Fx_{t-1} + G\eta_{t-1} \\ y_t &= h_t^T x_t + \xi_t \end{aligned} \quad j = 1, \ldots, R \qquad (7.17)$$

where y_t is the scalar stochastic observed variable, x_t is the state vector, $\eta_{j_{t-1}}$ is a k-dimensional vector of system disturbances, and ξ_j represents noise. $F \in \mathfrak{R}^{n \times n}$ and $h_t^T \in \mathfrak{R}^{1 \times n}$ are generally non-stochastic system matrices (Young et al. 1999; Young 2011).

Given the model derived from Equation (7.17), Kalman filtering can be used to estimate the space involved.

7.6.2 The Kalman Filter and Smoothing Algorithms

Dynamic harmonic regression estimates the time-varying parameters using a two-step (prediction-correction) Kalman filter, followed by a fixed-interval smoothing algorithm. The process for optimal eigenvector assignment can be described by the following equations adapted from Young et al. (1999), for each time period ($t = 1, ..., N$):

a. *Prediction*:

$$\hat{x}_{t|t-1} = F\hat{x}_{t-1}$$
$$P_{t|t-1} = FP_{t-1}F^T + GQG^T \tag{7.18}$$

b. *Correction*:

$$\Lambda_t = y_t - H\hat{x}_{t|t-1}$$
$$S_t = HP_{t|t-1} + R_t$$
$$K_t = P_{t|t-1}H^T S_t^{-1}$$
$$\hat{x}_{k+1|k+1} = \hat{x}_{k+1|k} + K_{k+1}\left(H_{k+1}P_{k+1|k}H_{k+1}^T + R_{k+1}\right)$$
$$\hat{x}_{k+1|k+1} = \hat{x}_{k+1|k} + K_{k+1}\left(H_{k+1}P_{k+1|k}H_{k+1}^T + R_{k+1}\right) \tag{7.19}$$

where $Q = diag\left(\sigma_v^2 \quad \sigma_\xi^2\right)$, with initial conditions x_o and x_o, and the notation $\hat{x}_{t|t-1}$ is used to indicate the estimate of $x(t)$ given the observations $x(0), ..., x(t-1)$. The recursive algorithm requires specifying the initial condition x_o and its error covariance P_o. In the formulation shown by equation 7.19, Λ_t denotes the one-step-ahead prediction-error (innovation), S_t is its variance, K_t is the optimal Kalman gain, and $\hat{x}_{t|t-1}$ and $P_{t|t-1}$ represent the updated state estimate and the updated estimated covariance for the state vector, respectively.

During the prediction state, the optimal estimate of the state vector at time t becomes

$$\hat{x}_{t|t-1} = F\hat{x}_{t-1|t-1} \tag{7.20}$$

and the associated covariance matrix of the estimation error is

$$P_{t|t-1} = FP_{t-1|t-1}F^T + GQG^T \tag{7.21}$$

The updating equations are then used to calculate a new estimate of the state as new observations arrive. The filter can be visualized as a feedback system, with the forward part driven by the innovations, which are a white noise sequence.

After the filtering stage, a fixed interval smoother is used to update (correct) the filter estimated state $\hat{x}_{t|t}$. In this case, using the output of the Kalman filter, smoothing takes the form of a backward recursion for $t = N, ..., 1$, operating from the end of the sample set to the beginning: Smoothing allows estimating the state vector at a given time point (given all the available data) and, hence, leads to interpolation of missing observations in the observed time series using the associated mean square error of the interpolated estimates. Different approaches can be used to avoid numerical issues.

Remarks:

- Equations (3.35) and (3.36) describe a recursive algorithm that runs backward in time from the last observation.
- In the first stage, the estimates $\hat{x}_{t|t-1}$ and $P_{t|t-1}$ are the one-step-ahead state estimates and the associated covariance matrix.

Compared to more traditional approaches, unobserved component models have the potential to include frequency information, local trends, and both oscillatory and irregular components.

7.7 Damage Detection

Damage or fault detection, as determined by changes in the modal properties of system response, is a subject that has received considerable attention in the literature. The basic idea is that modal parameters (mainly frequencies, mode shapes, and modal damping) are functions of the physical properties of the structure (mass, damping, and stiffness). Therefore, changes in the physical properties will cause changes in the modal properties.

Changes in modal properties, especially damping and modal frequency, are a key indication of global system health. Huang et al. (2010) discuss the practical implementation of these approaches for power system oscillatory monitoring in Figure 7.8.

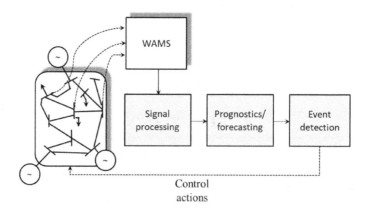

FIGURE 7.8
Envisioned prognostics and decision support system.

Experience suggests that the forecasting methods that have the better performance are those that explicitly model the trend and oscillatory components (Spyros et al. 2000).

7.8 Power Systems Time Series Forecasting

Short-term electric load forecasting

Mohan and Kumar (2018) provide a summary of state-of-the-art short-term electric load forecasting models. Examples include forecasting wind behavior and oscillatory behavior.

Short-term prediction of wind output

In Zavala and Messina (2014) a strategy for short-term prediction of wind output based on a window-based DHR approach was proposed. From the above developments, model prediction for wind power output at time t is given by

$$x_j(t) = F_j x_j(t-1) \qquad (7.22)$$

for $t = 1,\ldots,T$. A short-term prediction technique based on the use of Kalman filtering techniques can be developed to estimate wind power output using a sliding window approach. A conceptual representation of this model is shown in Figure 7.9.

To provide efficient short-term prediction, the method is applied to short segments of data. In this case, a moving data window is used. First, the size of each window, N_y, and the forecasting horizon F_h are defined; then, the number of moving samples N_w is selected. The parameters k^i and p^i are used to indicate where the *i*th window starts and finishes.

Formally, the model can be described as

$$\begin{aligned} k^i &= N_w(i-1)+1 \\ p^i &= N_s + N_w(i-1), \qquad i=1,2,\ldots,N. \end{aligned} \qquad (7.23)$$

The sliding window approach starts from the left, attempting to begin with the oldest data.

When a new sampling segment enters into the data window, the older sample is removed: the key step of this approach is the selection of an appropriate parameter, N_w, that allows for the use of a flexible moving window dependent on the time that every new sample is recorded. Figure 7.9 illustrates this

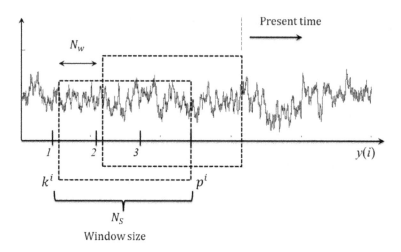

FIGURE 7.9
Sliding window approach. (Adapted from Zavala, A.J. and Messina, A.R., *Electr. Pow. Compo. Sys.*, 42, 1474–1483, 2014.)

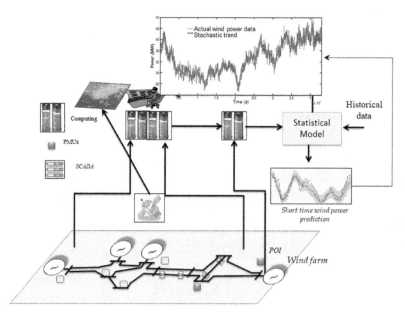

FIGURE 7.10
Schematic of the adopted forecasting technique. (Adapted from Zavala and Messina 2014.)

definition of data-adaptive windows—a new window occurs after period N_w, when a new measurement has been recorded. Figure 7.10 shows a schematic of a wind forecasting technique.

Extensions to this basic approach to allow a close-to-real-time implementation (e.g., $N_w = 1$) and typical sampling rates from 30 to 60 samples per second are being investigated. Note that, in this case, N_w and F_h are selected such that every new forecasting horizon starts where the previous window ends.

7.9 Anomaly Detection in Time Series

The problems of event (or anomalous) detection and signal segmentation can be approached from the perspective of time-series data mining (Agrawal and Agrawal 2015; Fakhrazari and Vakilzadian 2017; Esling and Agon 2012).

Anomaly detection is broadly defined as the process of finding the patterns in a data set whose behavior is neither normal nor expected. In this framework, dynamic event detection can be considered a precursor to fault diagnosis, which in turn, precedes prognosis. Depending on the data mining technique utilized, anomaly detection techniques be classified as clustering-based, classification-based, deep anomaly detection techniques, and hybrid approaches (Schwabacher and Kobel 2005). In addition, other automated procedures for dynamic event detection and location have been developed (Rovnyak and Mei 2011).

7.9.1 Extraction of Dynamic Trends

Trends are a useful first indication of slow changes occurring in the system. According to previous work in event detection, these methods can supplement data mining methods for event detection.

Several criteria to detect changes have been developed. These include:

- Energy tracking,
- Peak detection, and
- Cluster analysis.

In Messina (2015), a procedure to estimate changes in the observed behavior was developed based on the notion of local mean and the calculation of entropy. Other measures of a signal's strength include relative amplitude, frequency weighted amplitude, entropy, and norm. This section explores other approaches to event detection based on simple criteria.

Assume that a time series, y_t, is modeled as a linear combination of a time-varying mean, quasi-oscillatory components, and noise as

$$x(t) = x_t = m(t) + S(t) + e(t) \tag{7.24}$$

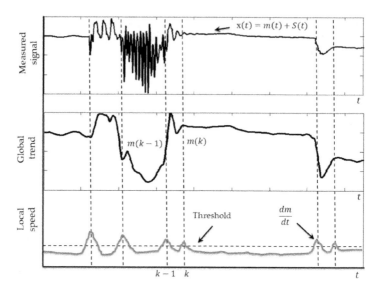

FIGURE 7.11
Schematic showing the use of local speed to determine anomalous events. Several variations to this formulation are discussed below.

where y_t is the observed time series, t denotes times, and $m(t), S(t), e(t)$ represent the slowly varying trend, quasi-cyclical components, and noise, respectively.

An illustration of this model is given in Figure 7.11. Visual inspection of this plot suggests that the slow time-varying trend can be used to segment the observed time series into homogeneous time intervals that are locally stationary. It is also observed that changes in a signal's behavior can be associated with a change in its trend.

7.9.2 Second Derivative Criterion

Conceptually, an abrupt change in the observed system behavior occurs when the first derivative of the trend function has a discontinuity (Matyasovszky 2011).

As illustrated in Figure 7.11, the first derivative identifies the zero crossings of the signal. Given a signal $x(t)$, an estimate of the first derivative can be obtained by analyzing the difference between $x(t)$ and $x(t-1)$.

With reference to Equation (7.24), the mean speed can be defined at time k as

$$\frac{dm(k)}{dt} = \frac{m(k) - m(k-1)}{t(k) - t(k-1)} \tag{7.25}$$

Using concepts from calculus, the second derivative criterion can be used to determine the peak of the oscillations. It follows from Equation (7.25) that the second derivative $m''(t)$ can be written as

$$m''(t) = \frac{dm^2(t)}{dt^2} = \frac{\dfrac{s(t)-s(t-1)}{t(k)-t(k-1)} - \dfrac{s(t-1)-s(t-2)}{(t-1)-(t-2)}}{t-(t-1)} \quad (7.26)$$

or

$$\frac{d^2m(t)}{dt^2} = \frac{s(t)-2s(t-1)+s(t-2)}{\Delta t^2} \quad (7.27)$$

where it has been assumed that $\Delta t = (t-1)-(t-2)=t-(t-1)$. These techniques can be applied to a wide range of problems in a near real-time setting including the identification of trends and changes in system prediction.

Table 7.2 summarizes three-point first -and second-order estimates for near-real time applications. These approximations are suitable for application with real-time signal processing techniques described in this chapter. When combined with a suitable data mining technique, these approaches can be used for power system data segmentation and event detection.

Figure 7.12 illustrates the application of the above approach to the computation of the second derivative of the instantaneous distance between two signals. Several application are envisaged, including:

- Detection of voltage instability and the formation of critical islands, and
- Monitoring of voltage angle deviations

Other relevant applications are described in Dosiek (2016).

Breaks, peaks, and slope changes may signal different physical regimes. Using these approaches, the time series can be converted into nearly stationary time-sequences of temporal behavior. Time interval pattern mining can then be used to analyze the quasi-stationary segments.

To illustrate, consider the frequency signal in Figure 7.13.

TABLE 7.2
Signal Trend

Method	Numerical Estimate
Three-point first derivative	$x'^{(k)} = x(k) - x(k-2)$
Second-order derivative	$x''(k) = x(k) - 2x1 + x(k-2)$

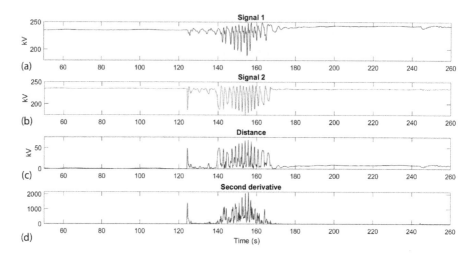

FIGURE 7.12
Illustration of distance and second derivative of a signal. a) Signal 1; b) Signal 2; c) Distance;
d) Second derivative.

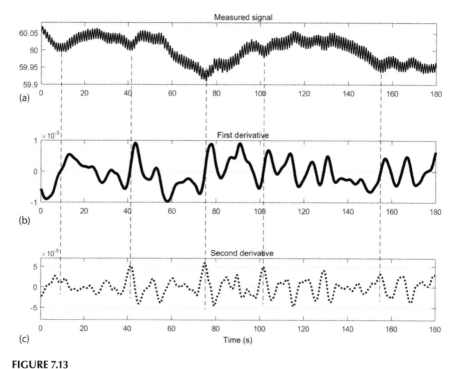

FIGURE 7.13
(a–c) Measured data with automatically detected peaks and valleys using the first/and second
derivative criterion.

Two comments are in order:

1. As shown in Figure 7.13, the second derivative reaches a maximum each time the measured signal reaches a minimum.
2. Each peak in the second derivative criterion defines the beginning or end of a time interval and the process is equivalent to time segmentation. Using this framework, the time-varying mean is systematically extracted and used for global system monitoring.

Post-processing of the local mean acceleration (second derivative) can be used to further improved event detection. Potential applications of these methods include trajectory (sequence) segmentation in which a trajectory is divided into a number of segments showing a common behavior.

References

Agrawal, S., Agrawal, J., Survey of anomaly detection using data mining techniques, *Procedia Computer Science*, 60, 708–713, 2015.

Begovic, M., Novosel, D., Karlsson, D., Henville, C., Michel, G., Wide-area protection and emergency control, *Proceedings of the IEEE*, 93(5), 876–891, 2005.

Borlase, S., *Smart Grids Infrastructure*, Technology and Solutions, CRC Press, Boca Raton, FL, 2012.

Callan, R., Larder, B., Sandiford, J., An integrated approach to the development of an intelligent prognostic health management system, 2006 *IEEE Aerospace Conference*, Big Sky, MT, March 2006.

Chow, S. M., Hamaker, E. L., Fujita, F., Boker, S. M., Representing time-varying cyclic dynamics using multiple-subject state-space models, *British Journal and Statistical Psychology*, 62, 683–716, 2009.

Dosiek, L., A framework for assessing methods of deriving frequency data from PMU voltage angles in the presence of measurement noise, *2016 North American Power Symposium*, Denver, CO, September 2016.

Esling, P., Agon, C., Time-series data mining, *ACM Computing Surveys*, 45(1), 12:1–12:34, 2012.

Fakhrazari, A., Vazilzadian, H., A survey on time series data mining, *2017 IEEE International Conference on Electro Information Technology (EIT)*, pp. 476–481, Lincoln, NE, May 2017.

Huang, Z., Diao, R., Zhou, N., Fuller, J. C., Tuffner, F. K., Mittelstadt, W. A., Chen, Y., Hauer, J. F., Trudnowski, D. J., Dagle, J. E., MANGO: Modal analysis for grid operation: A method for damping improvement through operating point adjustment, U.S. Department of Energy, Pacific Northwest Laboratory, Technical report PNNL-19890, Richland, WA, October 2010.

Hyndman, R. J., Khandakar, Y., Automatic time series forecasting: The forecast package for R, *Journal of Statistical Software*, 27(3), 1–22, 2008.

Ji, K. H., Herring, T. A., A method for detecting transient signals in GPS position time-series: Smoothing and principal component analysis, *Geophysical Journal International*, 193, 171–186, 2013.

Kalman, R. E., A new approach to linear filtering and prediction problems, *ASME Transactions Journal Basic Engineering*, 83-D, 3–14, 1960.

Kotu, V., Deshpande, B., Time series data mining, in *Data Science: Concepts and Practice*, pp. 395–445, 2nd ed., Elsevier, Cambridge, 2019.

Ljung, L., *System Identification*, 2nd ed., PTR Prentice Hall, Upper Saddle River, NJ, 1999.

Marple, S. L., *Digital Spectral Analysis with Applications*, Prentice Hall, Englewood Cliffs, NJ, 1987.

Martinelli, R., Rhoads, N., Predicting market data using the Kalman filter, *Technical Analysis of Stocks & Commodities Magazine*, 28(1), 44–47, February 2010.

Martinez, E., Messina, A. R., Modal analysis of measured inter/area oscillations in the Mexican interconnected system: The July 31, 2008 event, *2011 IEEE Power and Energy Society General Meeting*, Detroit, MI, July 2011.

Matyasovszky, I., Detecting abrupt climate changes on different time scales, *Theoretical and Applied Climatology*, 105(3), 445–454, 2011.

Messina, A. R. Wide-area monitoring of interconnected power systems, *IET Power and Energy Series*, Stevenage, UK, 2015.

Mohan, N., Kumar, S. S., A data-driven strategy for short-term electric load forecasting using dynamic mode decomposition model, *Applied Energy*, 232, 229–244, 2018.

Montgomery, D. C., Johnson, L. A., Gardiner, J. S., *Forecasting & Time Series Analysis*, McGraw-Hill, New York, 1990.

National Academies of Sciences, Engineering, Medicine, A consensus study report, in-time aviation safety management: Challenges and research for an evolving aviation system, April 2018.

North American Electric Reliability Corporation (NERC), *Real-Time Application of Synchrophasors for Improving Reliability*, Princeton, NJ, 2010.

Poslad, S., Middleton, S. E., Chaves, F., Tao, R. Necmioglu, O., Bugel, A. R., A semantic IoT early warning system for natural environment crisis management, *IEEE Transactions on Emerging Topics in Computing*, 3(2), 246–257, 2015.

Rovnyak, S. M., Mei, K., Dynamic event detection and location using wide area phasor measurements, *European Transactions on Electrical Power*, 21, 1589–1599, 2011.

Schwabacher, M., Goebel, K., A survey of artificial intelligence for prognostics, *AAAI Fall Symposium: Artificial Intelligence for Prognostics*, pp. 107–114, Arlington, VA, 2007.

Schwabacher, M., Oza, N., Matthews, B., Unsupervised anomaly detection for liquid-fueled rocket propulsion health monitoring, *Journal of Aerospace Computing, Information and Communication*, 6, 464–482, 2009.

Simo, S., *Bayesian Filtering and Smoothing*, Cambridge University Press, New York, 2014.

Sohn, H., Farrar, C., Hunter, N., Worden, K., Applying the LANL statistical pattern recognition paradigma for structural health monitoring to data from a surface-effect fast patrol boat, Los Alamos National Laboratory, Report LA-13761-MS, Los Alamos, New Mexico, 2001.

Spyros, M., Michele, H., The M3-competition: Results, conclusions and implications, *The International Journal of Forecasting*, 16, 451–47, 2000.

Xia, T., Dong, Y., Xiao, L., Du, S., Pan, E., Xi, L., Recent advances in prognostics and health management for advanced manufacturing paradigm, *Reliability Engineering and System Safety*, 178, 255–268, 2018.

Young, P. C., Pedregal, D. J., Tych, W., Dynamic harmonic regression, *Journal of Forecasting*, 18, 369–394, 1999.

Young, P. C., *Recursive Estimation and Time-Series Analysis: An Introduction for the Student and Practitioner*, 2nd ed., Springer-Verlag, Berlin, Germany, 2011.

Zavala, A. J., Messina, A. R., A dynamic harmonic regression approach to power system modal identification and prediction. *Electric Power Components and Systems*, 42(13), 1474–1483, 2014.

Section III

Challenges and Opportunities in the Application of Data Mining and Data Fusion Techniques

8

Data Fusion and Data Mining
Analysis and Visualization

8.1 Introduction

Multidimensional data visualization is a key task of modern data fusion and data mining analysis methods (Dzemyda et al. 2013, Sacha et al. 2017). Visualization methods are essential for studying and understanding the behavior and structure of large-scale data sets. They support decision-making activities and are a critical component of a prognostics and decision support system. In addition, actionable intelligence and decision-making requires effective visualization of dynamic data. Effective visualization of system dynamics is also critical to the analysis of wide-area situational awareness (NERC 2010).

These methods can also be useful to reveal hidden trends, highlight outliers, show clusters, and expose both gaps and other abnormalities in data. Other considerations include the detection and isolation of the root cause and finding the paths along which the disturbance propagates. This requires an understanding of the cause-and-effect relationships between multiple variables.

When combined with advanced data fusion techniques and appropriate metrics, visualization techniques also have the potential to increase situational awareness and aid decision support (Treinish 2001). In this chapter, an analysis framework based on projection methods for visualization and clustering of power system data is introduced.

Figure 8.1 gives a general framework for data visualization. Three main loops are considered: (a) A direct loop involving direct visualization of measured data, (b) An indirect loop involving data fusion and dimensionality reduction, and (c) An outer loop (dashed lines) that describes the interaction of the operator with the visualization techniques and enables model validation and assessment of mitigation measures. Data fusion may incorporate other techniques such as correlation analysis and data mining.

In the direct loop, (raw) data is directly visualized in near real-time (Zhang et al. 2007; Bank et al. 2007). Visualization techniques, on the other hand,

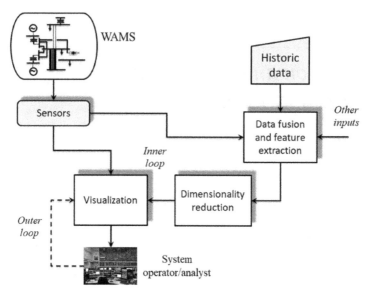

FIGURE 8.1
General framework for data visualization including an analyst-in-the-loop interface.

must support user interaction. An external loop (dashed line), where the operator or operating analyst can modify or visualize dynamic events, is also included.

In the succeeding sections, methods for data visualization and characterization of observational data are examined in the context of modern smart monitoring systems. Recent trends in data visualization are also examined.

8.2 Advanced Visualization Techniques

Dynamic trajectories collected by PMUs and other dynamic recorders contain substantial information about the dynamic behavior of the system. Organizing and condensing this information into a form that can provide useful understanding of system behavior during a perturbation is a challenge that is receiving increasing attention.

Measured data following system perturbations is time-varying in nature, which makes feature extraction and visualization difficult (Dutta and Overbye 2014). This is especially true in the case of propagating phenomena over large system regions. State of the art methods of visualizing propagating phenomena have primarily focused on two- and three-dimensional static representations of data. Much less attention has focused on situational awareness visualization of propagating faults.

There is a large and growing literature on data visualization including several books and scientific papers. Several visualization techniques have been developed and applied to represent data in various fields. The most common multi-dimensional visualization techniques include scatter plot matrices, two- and three-dimensional coordinate plots, heat and color maps, and heat-maps to mention a few approaches.

Table 8.1 summarizes some common visualization methods used across various disciplines from environmental analysis to power system applications. These approaches provide different perspectives on system behavior and are often used in a complementary manner.

Traditional 2D and 3D scatter plots capture dominant behavior and are often used to visualize static clusters. Scatter plots visualize multidimensional data by mapping two dimensions to the X and Y coordinates and offer a convenient way to determine the presence of linear correlations between multiple variables as well as to visualize each cluster individually.

Classical static maps, however, are limited in their ability to capture dynamic information. Among the various data mining visualization techniques, scatter plots have been used extensively to explore relationships between coordinates or variables. But, visualizing the data with individual scatter plots cannot show the development of the two variables over time.

The use of more advanced techniques for feature extraction and visualization is covered in Sections 8.2.1 through 8.2.4.

TABLE 8.1

Multidimensional Data Visualization Techniques

Technique	Approaches
Geospatial and spatiotemporal visualization methods	• Heat maps (Weber and Overbye 2000) • Plots of coordinates as a function of time (Liu et al. 2017) • Movie-like representations
Geometric methods	• Histograms, scatter plots, Andrews plots, etc. • Temporal cluster graphs (Adomavicius and Bockstedt 2008) • Dendograms (Juarez et al. 2006) • Matrix scatter plots • Three-dimensional scatter plot techniques (Dzemyda et al. 2013), etc.
Projection methods	• Linear projection methods • Nonlinear projection methods
Time-domain modal decomposition methods	• Koopman mode analysis • Dynamic mode decomposition, …

8.2.1 Geospatial and Spatiotemporal Maps

In the last few years, geospatial and spatiotemporal maps are being increasingly used to explore relationships within data sets, as well as to visualize spatiotemporal behavior.

Geospatial models offer a natural framework to display and explore spatial relationships within large data sets and can be used to visualize events over time. Further, geospatial information can be used as a part of an early warning system.

The starting point is a geographical representation, in which the physical location of sensors, transmission resources, and generation resources are displayed, as shown in Figure 8.2.

The procedure involves a transformation of measurement points (sensor locations) from geographic space in the power grid to points in the d-dimensional data space (a low-order representation). In this general set up, each measurement location is represented by a single point p in data space, and transformed back to geographic space (Hoffman et al. 2008) using a suitable transformation technique. A heat map representation can then be obtained showing relevant system behavior (Weber and Overbye 2000).

Conceptually, the analysis involves three main steps:

Step 1 *Data representation in physical space.* In this step, spatiotemporal data is represented by block matrices or tensors for a given time interval using the approaches shown in Chapters 3 and 4.

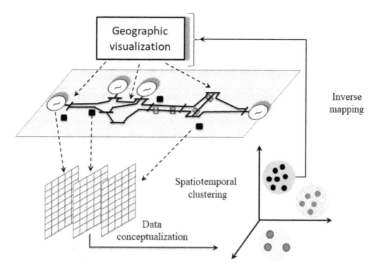

FIGURE 8.2
Overview of geospatial data visualization.

Step 2 *Data reduction.* Nonlinear projections of the data onto a low-dimensional space is a key step in visualizing multidimensional data. Ideally, the result is a low-dimensional representation (2 or 3 dimensions) in which relevant or hidden features are singled out. Alternatively, bio-inspired techniques could be used for visualization as discussed in Chapter 9. When combined with a sliding window approach, the method can be used to provide a near real-time representation of system behavior.

Step 3 *Data visualization in physical space.* This step involves transforming the low-dimensional data back to the original, physical (geographical) space; in order to correlate streaming wide-area information with the sensors' geographical information (Chai et al. 2016). Such a transformation is usually generated using an inverse mapping; successful use of this representation depends strongly on the ability to find coordinate transformations that correctly capture the relevant dynamics.

The first step in this process involves the representation of system measurements by matrices or tensors. Special cases of interest are:

1. The matrix/tensor representation is obtained for a given fault scenario. In this case, the clusters represent static behavior and can be obtained using standard clustering-based analysis techniques from Chapter 3. Visualization techniques such as those described in Chapters 4 through 6 can be used to visualize clusters

2. The multiblock block-dimensional matrix/tensor representation is associated with spatiotemporally evolving phenomena. Here, the emphasis is on visualization of fault propagation paths. To be of practical interest, the observed signals may be defined for a local time window as

$$\mathbf{X}^{TW_j} = \begin{bmatrix} x_1(t_{TW_1}) & x_2(t_{TW_1}) & \cdots & x_m(t_{TW_1}) \\ x_1(t_{TW_2}) & x_2(t_{TW_2}) & \cdots & x_m(t_{TW_2}) \\ \vdots & \vdots & \ddots & \vdots \\ x_1(t_{TW_N}) & x_2(t_{TW_N}) & \cdots & x_m(t_{TW_N}) \end{bmatrix}, j = 1, \ldots, T \qquad (8.1)$$

Analysis of this model results in time-varying clusters or energy propagation maps.

There are a few examples where the latter approach has been used to extract information from multidimensional trajectory data. Extensions and generalizations to this basic approach are discussed in Section 8.3.

8.2.2 Direct Visualization of Power System Data

Techniques to visualize system behavior from PMU data have been recently developed and integrated into several WAMS and situation awareness systems. These techniques collect and visualize data directly from sensors and therefore provide fast visualization of propagating phenomena—refer to Figure 8.1. Major examples include techniques to visualize wide area frequency and voltage propagating phenomena (Dutta and Obervye 2014). Similar efforts are also being developed for distribution systems (Donde and Mohamed 2016).

While these approaches can provide useful information about the time evolution of propagating phenomena, they lack important information needed for the design of remedial measures and the design of communication link infrastructures. Another limitation is that these techniques do not fuse data and, therefore, can only provide partial information about cause-effect relationships in the analysis of propagating phenomena.

8.2.3 Cluster-Based Visualization of Multidimensional Data

In this stage, a reduced-order representation is obtained in which dominant features are retained. As a by-product, clusters (feature-level) depicting overall system behavior are obtained. Finally, the extracted global behavior is mapped back to the original physical space for visualization.

Because dimensionality reduction results in low-dimensional models, efficient visualization techniques based on two- and three-dimensional representations are of interest (Chen et al. 2015).

Without loss of generality, assume that high-dimensional raw data is projected onto a low-dimensional space of dimension d for analysis and visualization. Formally, given the embedding $U \in \mathfrak{R}^d$

$$U : x \mapsto \{U_1, U_2, \ldots, U_d\} = \left\{ \begin{bmatrix} u_{11}^{(1)} \\ u_{21}^{(1)} \\ \vdots \\ u_{m1}^{(1)} \end{bmatrix}, \begin{bmatrix} u_{11}^{(2)} \\ u_{21}^{(2)} \\ \vdots \\ u_{m1}^{(2)} \end{bmatrix}, \ldots, \begin{bmatrix} u_{11}^{(d)} \\ u_{21}^{(d)} \\ \vdots \\ u_{m1}^{(d)} \end{bmatrix} \right\} \qquad (8.2)$$

where the mapping provides a realization of a graph G as a cloud of points in a lower-dimensional space. An alternative is the use of Fuzzy clustering techniques.

Recently, several data visualization techniques have appeared in the power systems literature mainly in connection with the application of linear and nonlinear projection methods and can be used to display clusters in a convenient form.

Example 8.1 examines the application of these techniques to mine and visualize clusters and further illustrates these ideas.

Example 8.1

As a specific example, speed deviations from transient stability simulations are used to visualize clusters using 2D and 3D visualization approaches. The data set is the same used in Figure 3.11 in Section 3.6 and in example 6.3. The aim is to visualize modal behavior using line and scatter plots.

First, the matrix U of eigenvectors (8.2) was determined using the algorithm in Table 7.1.

Figure 8.3a shows a plot of the three dominant eigenvectors of the Markov matrix, obtained using the Matlab function *gplotmatrix*—refer

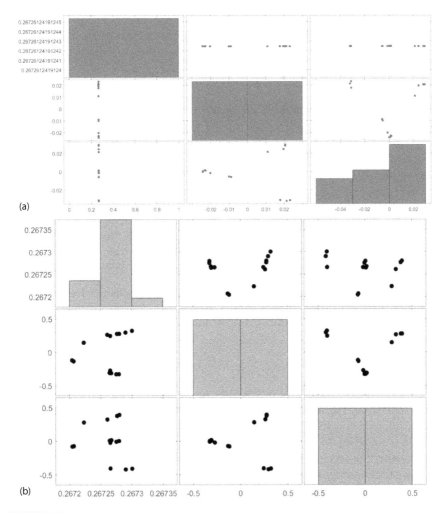

FIGURE 8.3
Scatter plot matrix of the matrix of dominant eigenvectors obtained using the Matlab command *gplotmatrix* (U,[]). (a) Diffusion Maps, (b) Markov clustering algorithm.

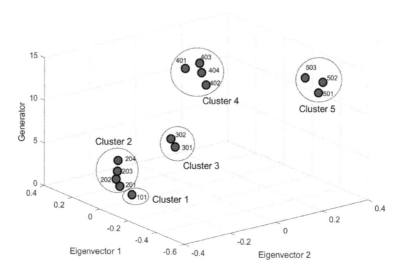

FIGURE 8.4
3D clustering of generator speed deviation measurements. Markov clustering algorithm.

to Figures 3.13 and 3.14b for a comparison with diffusion maps analysis. In this plot, the diagonal elements display the histograms. In the present context, the off-diagonal elements represent the projection of the 3D eigenvectors onto the 2D space.

For comparison, Figure 8.4 shows a three-dimensional plot of the main oscillatory modes determined using dynamic mode decomposition. *Gplotmatrix* representations and 3D representations provide complementary information on system behavior. In all cases, clusters are properly characterized.

Although visually appealing, these approaches however have the drawback that they provide only a static representation of the data. Chapter 10 will explore data driven health monitoring.

8.2.4 Moving Clusters

In Chapters 4 and 5, several techniques to visualize high-dimensional data were examined. These methods have proved useful for the visualization and extraction of a specific motion that is important for intersystem oscillations (Juarez et al. 2006). More general situations, however, would demand the use of trajectory clustering techniques to derive dynamic measures of system stability (Adomavicius et al. 2008).

As discussed in Cabrera et al. (2017), dynamic patterns may be affected by control actions, topology changes, or nonlinear effects. As a result, clusters evolve over time to match control actions and/or abrupt changes in system topology.

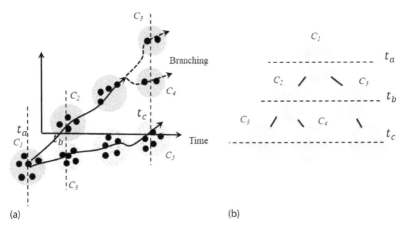

FIGURE 8.5
Time moving clusters illustrating cluster splitting and other characteristics. (a) Dynamic trajectories, and (b) Tree of clusters.

Several online hierarchical clustering methods based on pattern recognition techniques have been proposed for the automated clustering of system motion trajectories (Juarez et al. 2011). These techniques have applications in both operation and system monitoring.

Figure 8.5 shows an example of moving clusters showing branching of clusters (Chlis et al. 2017). Methodologies based on divide-and-merge techniques can be used for clustering dynamic trajectories (Cheng et al. 2005). Application of clustering techniques to dynamic trajectories creates trees that allow one to recursively partition dynamic trajectories into several subtrees as shown in the plot.

As suggested in the plot, analysis of the center of movement is critical to assess long-term system behavior. Intercluster distances can also be useful to facilitate comprehension of system separation mechanisms.

As discussed in Section 8.4 of this chapter, analysis of moving characteristics may be useful to analyze cases such as voltage collapse or system separation into islands (Figure 8.5).

8.3 Multivariable Modeling and Visualization

Intelligent monitoring and visualization of power system disturbances is a key step in the development and application of situational awareness algorithms (Cabrera et al. 2017).

These approaches can be used to integrate and process large amounts of multimodal, multi-type observational data from various monitoring technologies or data concentrators to assess the power system health in near-real time. Because of their strong analytical formulation, they can be used to determine communication needs, fuse multimodal data, and both detect and visualize interacting paths.

8.3.1 Formal Framework

Recently, a theoretically solid approach has been proposed that is suitable for analysis and visualization of power system disturbances (Cabrera et al. 2017). This approach has several advantages over conventional (two-block) PCA and PLS methods.

Following Messina (2015), consider a generic WAMS architecture consisting of M areas or control regions indexed $\{j = 1,\ldots,M\}$. To formalize the model, assume further that each area has a network of m_k, $\{k = 1,\ldots,m_k\}$ sensors deployed to monitor system behavior and let the time evolution of the measured signals be denoted by x_k.

An illustration of this concept for a multi-area system is given in Figure 8.6. Interest is focused on the analysis and visualization of propagating phenomena using data mining and data fusion techniques.

In developing the model, two basic assumptions are introduced:

1. The observed system response can be represented by a linear model, and

2. A distributed data fusion architecture is adopted.

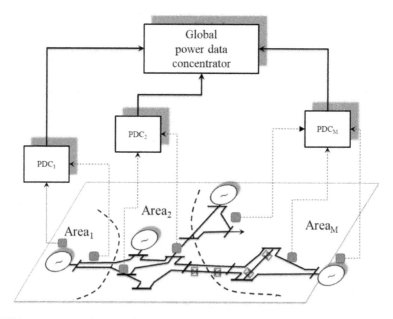

FIGURE 8.6
An illustration of a multi-area power system.

8.3.2 Multiblock Analysis Techniques

A natural extension of the framework introduced in the previous subsection is the joint analysis of multiscale data introduced in Chapter 5. As discussed in Chapters 1 and 2, measured data can be analyzed at the sensor or global (PDC) level.

In the distributed monitoring architecture, the set of local time histories at PDC level can be arranged into a raw-level observation matrix of the form:

$$X_j = \begin{bmatrix} x_1^j(t) \\ \vdots \\ x_{m_j}^j(t) \end{bmatrix} = \begin{bmatrix} x_1^j(t_1) & \cdots & x_1^j(t_N) \\ \vdots & \ddots & \vdots \\ x_{m_j}^j(t_1) & \cdots & x_{m_j}^j(t_N) \end{bmatrix} \in \Re^{m_j \times N}, \quad j = 1,\ldots,M \quad (8.3)$$

where each row represents the time evolution of measure data and each column denotes sensor location (Messina 2015; Cabrera et al. 2017). In developing the model, it is assumed that each block j has N rows (observations) and m_j columns (sensors). Moreover, each data block is associated with a specific area or control region. The MB-PCA method can be used to find a consensus between two or more blocks taken together.

Following the notation of Cabrera et al. (2017) and Messina (2015), the global observation matrix, X_g, can be defined as

$$X_g = \begin{bmatrix} X_1 & X_2 \ldots & X_M \end{bmatrix} \in \Re^{m \times N}, \quad M \geq 2 \quad (8.4)$$

with $m = m_1 + m_2 + \ldots + m_M$, where in general, data blocks may be of different dimensions (i.e., $m_i \neq m_j$); this results in a three-way decomposition of the data. Alternative representations can be generated by partitioning the data matrix into blocks of different dimensions, using vertical concatenation.

In interpreting this model, it should be observed that:

- Because of the potentially large dimension of the observation matrix, X_g, direct analysis may be prohibitive. Moreover, since each data block may be associated with a different physical variable, scaling techniques may have to be used to compare numerical results.

- The exploration of associations between the various data sets using two-block representations may not be possible.

Among the existing multiblock methods, consensus PCA (MB-PCA) and multiblock PLS can be used to construct PCA (PLS) models of high-dimensional power system data sets. A key advantage of these models is their simplicity and flexibility as well as their computability from single-block PCA and PLS factorizations and their ability to simultaneously describe phase and temporal variations. This is an issue that has not been discussed in the literature.

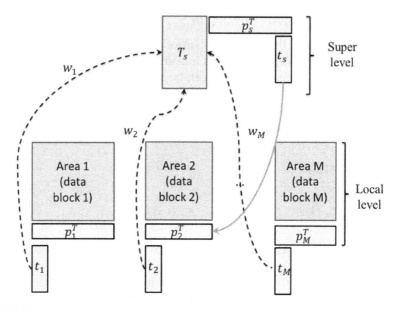

FIGURE 8.7
Multiblock PCA analysis. (Adapted from Westerhuis, J. A. et al., *J. Chemometr.*, 12, 301–321, 1998.)

Figure 8.7 shows an illustration of the adopted model for multiblock analysis adapted from Westerhuis et al. (1998). In this representation, data sets of varying dimensionality can be simultaneously analyzed and lead to a compact visualization of power system behavior. Each data block is assumed to represent different physical variables or similar data collected by different PDCs as suggested in Figure 8.6.

The consensus MB-PCA identifies a set of weights w_i that maximally capture the within-block or data variances as well as the between-the-block covariances. This decomposes the three-way array in the upper part of Figure 8.7 into a series of principal components consisting of score vectors t_j, loading matrices T_j, and a residual matrix E.

Upon decomposition, the global data matrix can be represented in the form

$$X_g = T_A P_A^T + E = \sum_{i=1}^{A} t_i p_i^T + E = \hat{X} + E \tag{8.5}$$

or, in terms of the notation used in Figure 8.7,

$$X = \begin{bmatrix} t_1 p_1^T & t_2 p_2^T \dots & t_M p_M^T \end{bmatrix} \tag{8.6}$$

where T_{sup} (again referring to Figure 8.7) is

$$T_t = \sum_{i=1}^{K} w_i t_i \qquad (8.7)$$

and similarly, the super weight vector, w_{sup}, is given by

$$w_t = \begin{bmatrix} w_1 & w_2 \dots & w_M \end{bmatrix}$$

Noting that

$$T_t = \begin{bmatrix} t_1 & t_2 \dots & t_M \end{bmatrix}$$
$$P_t = \begin{bmatrix} p_1 & p_2 \dots & p_M \end{bmatrix}$$

one has that

$$X = T_t P_t^T$$

Variations to this basic procedure are given in Westerhuis et al. (1998). With this approach, the importance of each block score can be measured by measuring the value of the score's block weight.

The adopted implementation of the method follows that of Xu and Goodacre (2012) and is briefly summarized in the chart Multiblock PCA Algorithm. Other approaches include CCA correlation analysis and multidimensional data fusion.

Multiblock PCA Algorithm

Given a set of data blocks $X_1, X_2, \dots, X_M$,
1. Choose an initial value t_{sup}
2. Compute p_i, and as

$$p_i = X_i^T \frac{t_{sup}}{\|X_i^T t_{sup}\|}, \quad i = 1, \dots, M$$

$$t_i = X_i p_i, \quad i = 1, \dots, M.$$

3. Set $T = \begin{bmatrix} t_1 & \cdots & t_M \end{bmatrix}$ and calculate

$$w = T^T t_{sup}$$
$$t_{sup} = T^T w$$
$$w = \frac{w}{\|t_{sup}\|}$$
$$p_i = X_i^T t_{sup}$$
$$E_i = X_i - t_{sup} p_i.$$

4. Set $E_i = X_i$, $i=1, \dots, M$, and $X = \begin{bmatrix} X_1 & \cdots & X_M \end{bmatrix}$. Return to step 1 and iterate until the algorithm converges.

The outcome is a global decomposition of the data in terms of loading and score matrices. A physical interpretation is as follows:

- The superscores of the consensus PCA, T_t are equivalent to those obtained with the conventional PCA analysis of the concatenated matrix of equation (8.4).
- Physically, the dynamic patterns between each pair of resulting scores give an approximation of the leading mode between the associated areas or data blocks. This allows several extensions to this approach that may be needed to analyze data from multiple sources as discussed later. Privacy concerns, however, may prevent utilities from directly sharing information (Taniar 2008).

A simple example is considered in Example 8.2 to illustrate the use of multi-block analysis techniques to fuse data.

Example 8.2

The system shown in Figure 8.8 is used to provide illustration of the analysis of simultaneous recordings. The data set being analyzed corresponds to frequency variations observed using PMUs following a loss-of-generation event and comprises a time series of frequency swings recorded at 18 points across the system using PMUs (Martinez and Messina 2011). Among the existing network PMUs measurements, 18 signals from three regional systems are selected for analysis; the type,

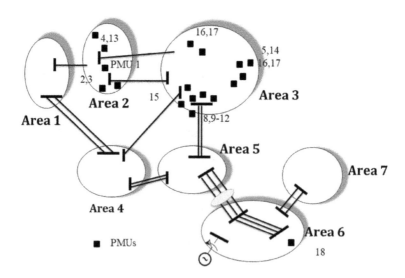

FIGURE 8.8

Schematic of the study system showing areas of interest and the location of installed PMUs. Measuring locations are displayed by filled black squares.

TABLE 8.2

Measurement Locations

PMU Number	Area	Sampling Rate (sps)[a]	Number of Sensors
1, 2, 3, 4, 13	2	20	5
5, 6, 7, 8, 9, 10, 11, 12, 14, 15, 16, 17	3	20	12
18	6	20	1

[a] Samples per second.

sampling rate, and sensor locations are given in Table 8.2. The observation set contains 3600 points and covers a period of 30 seconds.

Geographically, these measurements include three main regional systems: the signals are sampled for 250 seconds at a sampling rate of 20 samples per second and consist of 4900 data points. A summary of system events is provided in Martinez and Messina (2011). Figure 8.9 gives the time evolution of selected frequency time series of this event measured simultaneously at various system locations.

The three-block speed-based observation matrix is defined as $X_g = \begin{bmatrix} X_{A_2} & X_{A_3} & X_{A_6} \end{bmatrix} \in \mathfrak{R}^{N \times 18}$; attention is focused on the ability of the technique to characterize inter-area oscillations.

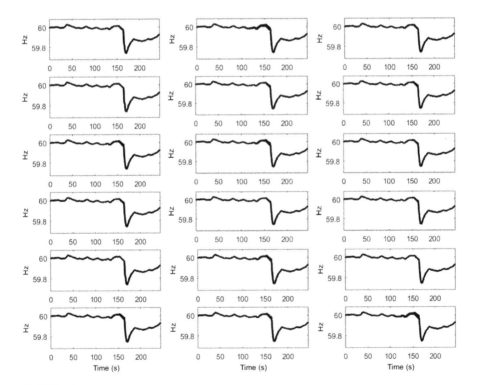

FIGURE 8.9
Time traces of measured data recorded simultaneously.

The feature vector is defined as

$$X_f = \begin{bmatrix} f_1 & f_2 \cdots & f_k \end{bmatrix} \in \mathfrak{R}^{N \times m_k} \tag{8.8}$$

where $f_1 = [f_j(t_1) \quad f_j(t_2) \cdots \quad f_j(t_N)]^T$, and $N=2404$, $m_2 = 5$, $m_3 = 12$, $m_6 = 1$.

The 2D visualization in Figure 8.10 shows that the first PC captures the slowest dynamics in the system. Following the procedure described in Section 8.3.1, the data matrix can be represented as

$$X_g = T_{sup} P_{sup}^T + E. \tag{8.9}$$

Comparing these values to that of the original signal, results in Figure 8.11.

The first principal component (PC 1) in Figure 8.12 corresponds to the overall measurements trend, while the second principal component can be associate with the unstable mode. These results are consistent with previous research using other modal decompositions.

8.3.3 Open Issues

The application of MB-PCA/PLS is sensitive to various factors including trends and missing data. When local measurements have different units, scaling may be required. This will be addressed in more detail in Chapter 10. A practical approach is to use nonlinear detrending.

8.3.4 Propagation Faults: Consensus PCA

Extensions of multiblock analysis techniques to analyze and visualize system disturbances have recently gained attention (Cabrera et al. 2017). Two basic objectives can be associated with the ability of the method to analyze and visualize power system disturbances: (a) To represent local and global behavior using consensus PCA, and (b) To take advantage of the spatial interpretation of the consensus model in terms of local and global coordinates. Here the process is briefly summarized—the reader is referred to (Cabrera et al. 2017) for further details.

Given a decentralized structure of the form in Figure 8.6, the basic idea behind the CPCA is to find a consensus direction among from the blocks (Westerhuis et al. 1998) that maximally capture block variations.

In the standard PCA, each block X_j can be decomposed in the form

$$X_j = \sum_{l=1}^{r} t_j^l p_r^T$$

where the above parameters have the usual interpretation.

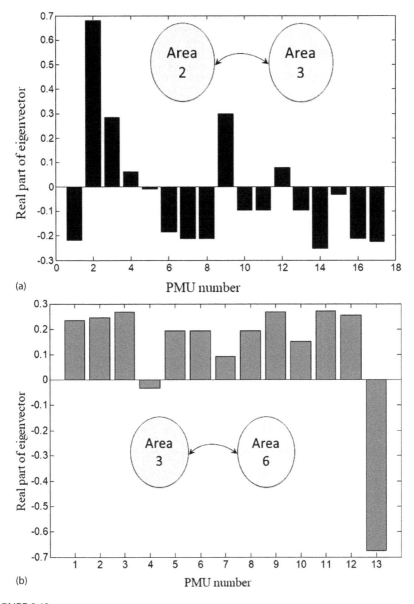

(a)

(b)

FIGURE 8.10
Spatial patterns extracted using (8.9). (a) Spatial pattern extracted from of T_{sup} in (8.9) associated with Areas 2 and 3. (b) Spatial pattern extracted from of T_{sup} in (8.9) associated with Areas 3 and 6.

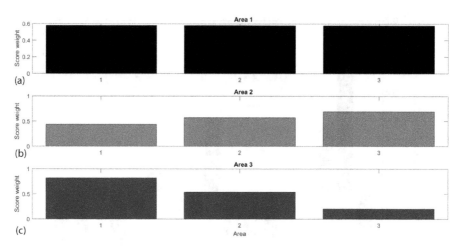

FIGURE 8.11
Score weights for Areas 1-3. a) Area 1; b) Area 2; c) Area 3.

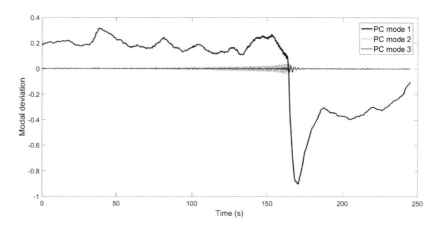

FIGURE 8.12
The three principal components for measured data in Figure 8.9.

Having computed the weights, the jth local observation matrix can be approximated by

$$
X_j = \begin{bmatrix} x_1^j(t) \\ \vdots \\ x_{m_j}^j(t) \end{bmatrix} = \begin{bmatrix} \varphi_1^j \hat{\varphi}_{11}^j \hat{a}_1(t) & \cdots & \varphi_r^j \hat{\varphi}_{1r}^j \hat{a}_1(t) \\ \vdots & \ddots & \vdots \\ \varphi_1^j \hat{\varphi}_{m_j1}^j \hat{a}_1(t) & \cdots & \varphi_r^j \hat{\varphi}_{m_jr}^j \hat{a}_1(t) \end{bmatrix} \in \mathfrak{R}^{m_j \times N}, \quad j = 1, \ldots, M
$$

$$(8.10)$$

and

$$
X_j =
\begin{bmatrix}
\hat{\varphi}_{11}^j & \cdots & \hat{\varphi}_{11}^j \\
\vdots & \ddots & \vdots \\
\hat{\varphi}_{11}^j & \cdots & \hat{\varphi}_{11}^j
\end{bmatrix}
\begin{bmatrix}
\varphi_1^j & \cdots & 0 \\
\vdots & \ddots & \vdots \\
\varphi_1^j & \cdots & \varphi_1^j
\end{bmatrix}
\begin{bmatrix}
\hat{a}_1(t) \\
\vdots \\
\hat{a}_1(t)
\end{bmatrix}
= \begin{bmatrix} \widetilde{\Phi}_1 & diag(\hat{\varphi}_1) & \hat{\Gamma}_g \end{bmatrix}, j = 1,\ldots, M
$$

(8.11)

This approach allows to examine relationships between control centers as well as to visualize the propagating energy following major events. Refer to Cabrera et al. (2017) for a detailed description of the application of these models to visualize propagating phenomena following power system disturbances.

Central to this approach, is its ability to exploit the distributed processing capacity of modern WAMS architectures to visualize the system dynamic behavior following disturbances. This is an issue that has received scarce attention in the power system literature.

8.3.5 Bandwidth of the Communication Networks

These frameworks can be used to assess the communication and computational support requirements of the distributed architecture. Open areas of research include how the areas should share the information responsibility to visualize global behavior, as well as communication and algorithmic needs (Cabrera et al. 2017).

8.4 Cluster-Based Visualization of Multidimensional Data

Inspection of Figure 8.2 suggests that the temporal evolution of cluster centroids can be used to analyze and visualize complex dynamic phenomena. This section explores the use of techniques to visualize the temporal behavior of measured signals using fuzzy clustering techniques.

8.4.1 Motivation

According to the description in Chapter 3, the centroid of a group of variables X can be determined by solving the optimization problem

$$
f(X_0) = \sum_{k=1}^{K} \sum_{i=1}^{n} u_{ki}^m \|x_i - C_k\|^2 , \text{ subject } t
$$

$$
\sum_{k=1}^{K} u_{ki} = 1, i = 1, \ldots, n
$$

(8.12)

$$
u_{ki} \geq 0, k = 1, \ldots, K, i = 1, \ldots, n
$$

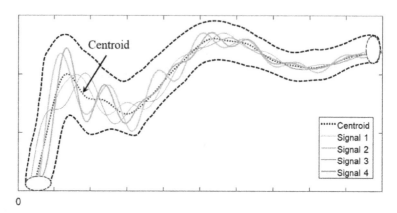

FIGURE 8.13
Visualization of *c*-means clusters. The dotted black curve is the centroid.

By segmenting the set of measurements, an approximation to the temporal clusters in Figure 8.13 can be obtained. Then, the distance between centroids can be used to determine other aspects of system behavior, such as the separation of the system into islands and cluster splitting.

Two practical applications of this approach are discussed in Subsections 8.4.2 and 8.4.3: voltage visualization and wide-area frequency visualization.

8.4.2 Wide-Area Voltage Visualization

A natural application of these ideas is the visualization of voltage control areas. Given voltage and phase angle measurements, voltage control areas can be determined using the following approach:

1. Select the number of voltage control areas based on physical or previous application of clustering-based approaches.
2. Using measurements of critical bus voltage magnitudes and voltage phase angles, unwrap the voltage phase angle at each measurement location and determine the measurement vector $X_\theta = [\theta_1 \quad \theta_2... \quad \theta_n]$. If necessary, also estimate phase angles at unmeasured system locations.
3. Partition the transmission buses into coherent groups using the *c*-means approach as

$$f(X_\theta) = \sum_{k=1}^{K} \sum_{i=1}^{n} u_{ki}^{m} \left\| x_i - C_k \right\|^2, \text{ subject } t$$

$$\sum_{k=1}^{K} u_{ki} = 1, \, i = 1, ..., n \qquad (8.13)$$

$$u_{ki} \geq 0, \, k = 1, ..., K, \, i = 1, ..., n$$

4. Compute the time evolution of the centroids and choose the dominant bus for area, k, $V_k, k = 1, \ldots, K$ such that

$$V_k, k = 1, \ldots, K.$$

5. Compute correlations between pilot buses and reactive power sources using CCA, PLSC, SVM, or other analysis techniques.

A flowchart which includes additional inputs to the data fusion scheme is shown in Figure 8.14. The appealing feature of this technique is its ability to jointly analyze bus voltage magnitude and reactive power signals.

Example 8.3

Here, the approach is applied to transient stability data from the Australian test system in Chapter 3 to find areas of coherent voltages. In this analysis, system dynamic behavior associated with data from regional systems was analyzed using partial least squares and PCA. The base case measurement vector is defined as $X = [X_\theta \quad X_V]$, with $X_\theta = [\theta_1 \quad \theta_2 \ldots \quad \theta_{59}]$, $X_V = [V_1 \quad V_2 \ldots \quad V_{59}]$ where the θ's represent the unwrapped bus voltage phase angles and the V's denote bus voltage magnitudes. Improved characterization can be obtained by augmenting the measurement vector with reactive power outputs as discussed in Figure 8.14.

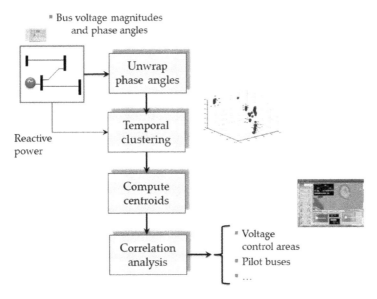

FIGURE 8.14
Flowchart for joint analysis of bus voltage magnitudes and phases (4.10).

TABLE 8.3

Clustering of Bus Voltage Control Areas

Cluster	Buses
1	413, 414, 415, 412, 411, 410, 401, 409
2	209, 210, 215, 207, 208
3	408, 402, 405, 404, 406, 407
4	205, 206, 201, 202, 204, 416
5	214, 213, 211, 216, 212, 217
6	508, 504, 501, 506, 503, 507, 509, 505, 502
7	102, 101, 309, 303, 301
8	315, 308, 306, 307, 313, 304, 314, 311, 305, 310, 312, 304

To illustrate the use of the proposed framework, 8 voltage control areas are considered. Table 8.3 shows bus voltage clusters determined using the c-means clustering algorithm in Chapter 3.

To illustrate the monitoring approach, let the simultaneous measurements of bus voltage magnitudes and reactive power be denoted by matrices X_V and X_Q. Assuming that X_V and X_Q are centered and normalized, the correlation matrix is computed as

$$R = X_V^T X_Q = U \Lambda V^T,$$

where U is the matrix of normalized eigenvectors of RR^T V is the matrix normalized eigenvectors of $R^T R$ is the diagonal matrix of singular values. Finally, as shown in Van Roon et al. (2014), the latent variables of X_V and X_Q are obtained by projecting the data onto their respective saliences

$$L_x = X_V V$$
$$L_y = X_Q U$$

8.4.3 Data Mining-Based Visualization Methods

The integration of data visualization techniques with change detection algorithms allows for a detailed characterization of propagating phenomena. In this section, a cluster-based visualization technique is proposed which can be used to examine various aspects of system behavior.

Two approaches to data-driven disturbance propagation phenomena are examined: (a) wide-area disturbance propagation, and (b) wide-area voltage propagation. These include wide-frequency visualization and disturbance propagation analysis and visualization (Pagnier et al. 2018; Lee et al. 2018).

Recent work has broadened the domain of problems considered. Major aspects of interest include:

- Propagation delay,
- System separation (cutset),
- Island formation, and
- Parameter deviation (maximum frequency deviations, etc.).

Section 8.5 examines the application of spatial and network displays to visualize system behavior.

8.5 Spatial and Network Displays

Recent work that examining correspondence between two very different types of data has shown promising results. Visualization of spatiotemporal data is of special interest to the development of situation awareness systems.

Let $X(t)$, denote a set of observations where N is time and m represents the number of sensors. Based on previous theory, assume that spatiotemporal behavior can be approximated by a model of the form

$$\hat{X} \approx \sum_{j=1}^{m} \hat{X}_j = \sum_{j=1}^{m} \phi_j a_j(t) = \Phi A(t) \tag{8.14}$$

where $\Phi = \begin{bmatrix} \phi_1 & \phi_2 \dots & \phi_r \end{bmatrix} \in \mathfrak{R}^{m \times r}$ is the transformation matrix containing the first m spatial modes, and $A(t) = \begin{bmatrix} a_1 & a_2 \dots & a_r \end{bmatrix}^T \in \mathfrak{R}^{m \times n}$ is the matrix of temporal coefficients, where $a_j(t) = \begin{bmatrix} a_j(t_o) & a_j(t_1) \dots & a_j(t_N) \end{bmatrix}$ are the vectors of the temporal coefficients. As discussed, examples of these representations include the proper orthogonal decomposition method, Koopman (DMD) models, and DMs, to mention a few methods.

Models of the form derived from Equation (8.3) have a physical interpretation of interest:

- Each term of the form $\phi_j a_j(t)$ can be interpreted in terms of sub-blocks (matrices) composed of the product of two tensors as illustrated in Figure 8.15.
- These representations can be considered as the unfolding of spatio-temporal data.

Typical applications of tensor decompositions are the analysis of time-varying graphs, where each slice of the tensor represents one snapshot of the graph (Lahat et al. 2015).

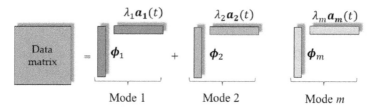

FIGURE 8.15
Matrix representation framework for models of the form 8.14.

Example 8.4

In this example, the data representation from Example 4.1 is used to obtain matrix-based modal representations. Starting from a modal decomposition of the form

$$\hat{X} \approx X_1 + X_2 + X_3 + X_4 + X_5 = \sum_{i=1}^{5} \underbrace{t_i p_i^T}_{X_i}. \tag{8.15}$$

Having obtained the model from Equation (8.8), the distance matrices D_i were computed using the approach from Chapter 3. It is noted that other representations could be obtained.

Figure 8.16 display energy maps for the matrices X_1 and X_2. Such representations provide a measure of interactions or relationships between states at a given scale and may be used to help diagnose modal interactions, extract sparsity patterns for major modes, and derive insight from spatiotemporal data.

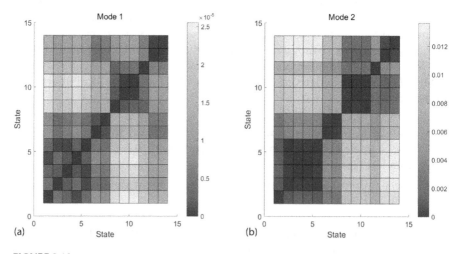

FIGURE 8.16
Surface plot of two major modal components in (8.15). (a) Surface plot of matrix X_1, and (b) Surface plot of matrix D_2.

When combined with multiblock representations, it is possible to compute, and visualize interrelationships between modes using physical space as well as to infer the presence of spatial features. Extensions to derive tensor factorization representations of the data are described in Afshar et al. 2017.

References

Adomavicius, G., Bockstedt, J., C-trend: Temporal cluster graphs for identifying and visualizing trends in multiattribute transactional data, *IEEE Transactions on Knowledge and Data Engineering*, 20(6), 721–735, 2008.

Afshar A., Jo, J. C., Dilkina, B., Perros, I., Khalil, E. B., Xiong, L., Sunderam, V., CP-ORTHO: An orthogonal tensor-factorization framework for spatio-temporal data, *26th ACM SIGSPATIAL International Conference on Advances in Geographic Information Systems*, Redondo Beach, CA, November 2017.

Bank, J. N., Gardner, R. M., Tsai, S. J. S., Kook, K. S., Liu, Y., Visualization of wide-area frequency measurement information, *2007 IEEE Power Engineering Society General Meeting*, Tampa, FL, 2007.

Cabrera, I. R., Barocio, E., Betancourt, R. J., Messina, A. R., A semi-distributed energy-based framework for the analysis and visualization of power system disturbances, *Electric Power Systems Research*, 143, 339–346, 2017.

Chai, J., Liu, Y., Guo, J., Wu, L., Zhou, D., Yao, W., Liu, Y., King, T., Gracia, J. T., Patel, M., Wide-area measurement data analytics using FNET/GridEye: A review, *2016 Power Systems Computation Conference*, Genoa, Italy, June 2016.

Chen, X. C., Faghmous, J. H., Khandelwal, A., Kumar, V., Clustering dynamic spatio-temporal patterns in the presence of noise and missing data, *Proceedings of the Twenty-fourth International Conference on Artificial Intelligence (IJCAI 2015)*, pp. 2575–2581, IJCAI, Inc., 2015.

Cheng, D., Kannan, R., Vempala, S., Wang, G., A divide-and-merge methodology for clustering, *ACM Transactions on Database Systems*, 31(4), 1499–152, 2005.

Chlis, N. K., Wolf, F. A., Theis, F. J., Model-based branching detection in single-cell data by K-branches clustering, *Bioinformatics*, 33(20), 3211–3219, 2017.

Donde, V., Mohamed, S., Data fusion and analytics applications for PG&E's power distribution systems, i-PCGRID workshop, https://ipcgrid.ece.msstate.edu/presentations/2016/, Mississippi State University, 2016.

Dutta, S., Overbye, T. J., Feature extraction and visualization of power system transient stability results, *IEEE Transaction on Power Systems*, 29(2), 966–973, 2014.

Dzemyda, G., Kurasova, O., Zilinskas, J., *Multidimensional Data Visualization: Methods and Applications*, Springer, New York, 2013.

Hoffman, F. M., Hargrove, W. W., Mills, R. T., Mahajan, S., Erickson, D. J., Oglesby, R. J., Multivariate spatio temporal clustering (MSTC) as a data mining tool for environmental applications, iEMS 2008, *International Congress on Environmental Modelling and Software*, Barcelona, Spain, 2008.

Juarez, C., Messina, A. R., Ruiz-Vega, D., Analysis and control of the inter-area mode phenomenon using selective One-Machine Infinite Bus dynamic equivalents, *Electric Power Systems Research*, 76(4), 180–193, 2006.

Lahat, D., Adali, T., Jutten, C., Multimodal data fusion: An overview of methods, challenges and prospects, *Proceedings of the IEEE*, 103(9), 1449–1477, 2015.

Lee, H. W., Zhang, J., Modiano, E., Data-driven localization and estimation of disturbance in the interconnected power system, ArXiv preprint, arXiv:1806.01318, 2018.

Liu, S., Maljovec, D., Wang, B., Bremer, P. T., Pascucci, V., Visualizing high-dimensional data: Advances in the past decade, *IEEE Transactions on Visualization and Computer Graphics*, 23(3), 1249–1268, 2017.

Martinez, E., Messina, A. R., Modal analysis of measured inter/area oscillations in the Mexican interconnected system: The July 31, 2008 event, *2011 IEEE Power and Energy Society General Meeting*, Detroit, MI, July 2011.

Messina, A. R., *Wide-Area Monitoring of Interconnected Power Systems*, IET Power and Energy Series 77, London, UK, 2015.

North American Electric Reliability Corporation (NERC), *Real-Time Application of Synchrophasors for Improving Reliability*, Princeton, NJ, 2010.

Pagnier, L., Jacquod, P., Disturbance propagation, inertia location and slow modes in large-scale high voltage power grids, ArXiv preprint, ArXiv:1810-04982, 2018.

Sacha, D., Zhang, L., Sedlmair, M., Lee, J. A., Peltonen, J., Weiskopf, D., North, S., Keim, D. A., Visual interaction with dimensionality reduction: A structured literature analysis, *IEEETransactions on Visualization and Computer Graphics*, 23(1), 241–250, 2017.

Taniar, D., *Data Mining and Data Knowledge Discovery Technologies*, IGI Publishing, Covent Garden, London, UK, 2008.

Treinish, L. A., Visual data fusion for decision support applications of numerical weather prediction, *17th International Conference on Interactive Information and Processing Systems for Metereology*, Oceanography, and Hydrology (IIPS), AMS, Albuquerque, New Mexico, 2001.

Van Roon, P., Zakizadeh, J., Chartier, S., Partial least squares tutorial for analyzing neuroimaging data, *The Quantitative Methods for Psychology*, 10(2), 200–215, 2014.

Weber, D., Overbye, J., Voltage contours for power system visualization, *IEEE Transaction Power System*, 15(1), 404–409, 2000.

Westerhuis, J. A., Kourti, T., Macgregor, J, F., Analysis of multiblock and hierarchical PCA and PLS models, *Journal of Chemometrics*, 12, 301–321, 1998.

Xu, Y., Goodacre, R., Multiblock principal component analysis: An efficient tool for analyzing metabolomics data which contain two influential factors, *Metabolonomics*, 8, S37–S51, 2012.

Zhang, G., Hirsch, P., Lee, S., Wide area frequency visualization using smart client technology, *2007 IEEE Power Engineering Society General Meeting*, Tampa, FL, 2007.

9

Emerging Topics in Data Mining and Data Fusion

9.1 Introduction

In recent years, data mining and data fusion techniques have been attracting a large amount of attention in a wide variety of applications such as knowledge discovery (Taniar 2008), situational awareness visualization (Ghamisi et al. 2019), spatiotemporal data mining (Shekar et al. 2015), and decision making (Bisantz et al.1999). However, there are several issues that make the development of data mining and data fusion techniques an extremely difficult and challenging task. Several areas have been identified as being of primary importance for future data mining and data fusion techniques.

Emerging topics in data fusion and data mining include:

- Dynamic spatiotemporal clustering;
- Analysis of distributed data arising from distribution sensor networks;
- Effective storage and indexing of massive spatiotemporal data;
- Pre-processing and denoising technology to address spatiotemporal data uncertainty;
- Robust spatiotemporal pattern mining and prediction algorithms;
- Spatiotemporal data visualization;
- The discovery of special associations and spatiotemporal events;
- Situational awareness visualization;
- Visual data mining;
- Development of sparse, low-rank and subspace fusion approaches to address the high dimensionality of features; and
- Classification and fusion of large-scale data.

In this chapter, an overview of recent advances in modeling spatiotemporal data is provided. Emerging research topics and techniques that are expected to be critical in the development of next-generation data mining and data fusion methods are described, with emphasis placed on applications to high dimensional power system data.

Challenges in data fusion and data mining are identified and discussed and new directions are examined. Potential future directions and challenges of data fusion integration are also discussed.

9.2 Dynamic Spatiotemporal Modeling

Wide-area PMU measurement data is inherently spatiotemporal in nature (Atluri et al. 2017). Although significant progress has been made in the development of true spatiotemporal models, capturing wide-area dynamic behavior using a limited number of sensors remains a challenging task.

Existing approaches to wide-area monitoring suffer from a number of limitations:

- Conventional data mining and data fusion techniques provide information about a dynamic process at a fixed point of time or study window. This prevents their application in a near real-time forecast system.

- Many techniques are matrix-based which makes impractical for the analysis of high-dimensional data.

- Traditional spatial clustering techniques are based on static approaches or are extracted using a fixed time. As a result, these approaches may fail to capture kinetic information such as branching and bifurcations and other characteristics and may change in size and other statistical properties over time (Chen et al. 2015; Kisilevich et al. 2010).

- Markov state models offer an alternative to incorporate kinetic information using simple models.

Analysis of these issues will require fundamentally new techniques for analysis. These issues are briefly discussed in Sections 9.2.1 and 9.2.2.

9.2.1 Tensors for Data Mining and Data Fusion

The vast majority of data fusion techniques that work with measured data are based on matrix representations. Since these matrices are dense, these methods are limited in their ability to analyze large-scale data and are prone to higher computational cost.

Recent studies have investigated the use of tensor and matrix decompositions for data mining and data fusion of multiblock representations (Acar et al. 2011; Papalexakis et al. 2016; Kolda and Sun 2008). Using this approach, the problem of joint analysis of data from multiple sources can be cast as a coupled matrix and tensor factorization problem (Acar et al. 2011; Papalexakis et al. 2016; Kolda and Bader 2009). These methods have the potential to complement conventional multiblock analysis by presenting a more detailed analysis of temporal behavior.

Consider, to illustrate these ideas, a distributed architecture consisting of L control areas or utilities. Assume further that each area has a network of m_k, $k = 1, \dots, L$ sensors and that this information is collected at a regional PDC in which a data fusion center is assumed to be integrated. Figure 9.1 is a simplified conceptual schematic of a multi-area power system which includes regional PDCs embedded in a WAMS.

Following the same notation as Messina (2015), let the observed multichannel signal $X \in \mathfrak{R}^{N \times j}$ at the jth sensor be given by a model of the form (4.1). Conventionally, the local observation matrices X_i^j, coming from the various PDCS are converted to a two-dimensional data matrix by unfolding (matricizing) the data. Let the kth observation matrix be defined as PDC

$$X_1^j = \begin{bmatrix} x_1 & x_2 & \cdots & x_m \end{bmatrix} = \begin{bmatrix} x_1(t_1) & x_2(t_1) & \cdots & x_m(t_1) \\ x_1(t_2) & x_2(t_2) & \cdots & x_m(t_2) \\ \vdots & \vdots & \ddots & \vdots \\ x_1(t_N) & x_2(t_N) & \cdots & x_m(t_N) \end{bmatrix} \in \mathfrak{R}^{N \times m} \quad (9.1)$$

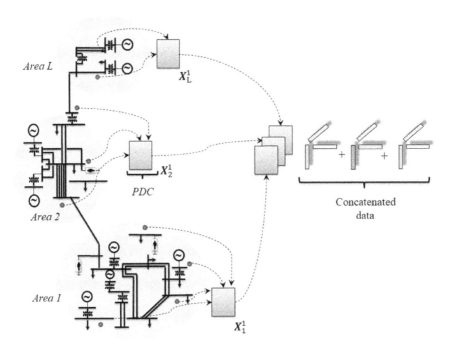

FIGURE 9.1
Conventional framework for power system monitoring.

A multiblock representation can then be obtained by horizontally concatenating the individual PDC measurements to yield a matrix

$$\hat{X} = \begin{bmatrix} \underbrace{X^1}_{PDC\,1} & \underbrace{X^2}_{PDC\,2} \cdots & \underbrace{X^L}_{PDC\,L} \end{bmatrix} \tag{9.2}$$

where each $X^1, k = 1, \ldots, L$, represents a slice of the representation (9.2). Here, the subscript denotes data collected at PDC k.

Specialized techniques based on multiblock principal component analysis have been devised to take advantage of the structure of this model. Two factors limit the applicability of these models:

- The large volumes of data, and
- The heterogeneous character of the individual observation matrices.

In classical PCA analysis, models of the form (9.2) can be analyzed multiblock analysis.

9.2.2 Tensors for Data Mining and Data Fusion

A second application of interest arises from the application of time-frequency analysis techniques to the analysis of a subset of modes (a frequency band). In this case, each measurement x_j is decomposed into its contributions in a given time-frequency band, and multiview analysis is applied to the resulting model.

For sensor j in area k, the measured response is expressed in the form

$$X_j^{m_k} = m_j(t) + \left(\sum_{k=1}^q A_k(t) \cos\left(\omega_k(t)\right) + \varphi_k(t) \right) \tag{9.3}$$

where $m_k(t)$ is the local temporal trend; $A_k(t)$, $\varphi_k(t)$ are the temporal amplitudes and phases, respectively; and q denotes the number of modes within a frequency band $\omega_{min} \le \omega_k \le \omega_{max}$. It should be observed that this model is general and could represent various time decomposition methods.

This information can be arranged into feature vectors $X_j^{m_k} = \begin{bmatrix} A_k(t_n) & A_k(t_n) \cdots & A_k(t_n) \end{bmatrix}^T, k = 1, \ldots, q$, and $n = 0, \ldots, n$, and used for state monitoring, assessment, and prediction. The feature vectors can then be collected into feature matrices of the form

$$X_j^{m_k} = \begin{bmatrix} x_1 & x_2 & \cdots & x_m \end{bmatrix} = \begin{bmatrix} A_1(t_1) & A_2(t_1) & \cdots & A_q(t_1) \\ A_1(t_2) & A_2(t_2) & \cdots & A_q(t_2) \\ \vdots & \vdots & \ddots & \vdots \\ A_1(t_N) & A_2(t_N) & \cdots & A_q(t_N) \end{bmatrix} \in \mathfrak{R}^{N \times q} \quad (9.4)$$

An illustration of this model is given in Figure 9.2. Similar to the previous development, each matrix $X_j^{m_k}$ can be seen as a slice of third-order tensor. The global measurement matrix then becomes

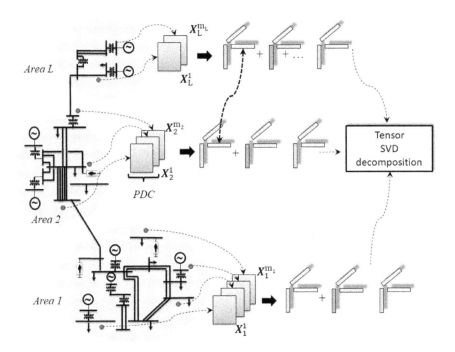

FIGURE 9.2
Tensor-based framework representation for power system monitoring. Note that each data matrix X_i^j can be considered as a slice of a third order tensor.

$$\widehat{X} = \left[\underbrace{X_1^1 \quad X_2^1 \ldots}_{PDC\,1} \quad \underbrace{X_1^1 \quad X_2^1 \ldots}_{PDC\,2} \quad \cdots \quad \underbrace{X_1^1 \quad X_2^1 \ldots}_{PDC\,L} \right] \tag{9.5}$$

The feature space described by these models is high-dimensional and sparse and it is amenable for tensor analysis.

The process can be summarized as follows: $X_j^{m_k}$

1. For each area k, decompose power system measured data into its contributions in a selected region of time-frequency space.
2. Construct the expanded data matrix.

The concatenated data matrix becomes

$$\widehat{X} = \left[\underbrace{X_1^1 \quad X_2^1 \ldots}_{PDC\,1} \quad \underbrace{X_1^1 \quad X_2^1 \ldots}_{PDC\,2} \quad \cdots \quad \underbrace{X_1^1 \quad X_2^1 \ldots}_{PDC\,L} \right] \tag{9.6}$$

3. Perform tensor analysis.

These models are amenable to tensor analysis. With reference to Figures 9.1 and 9.2, a tensor matrix factorization for each regional PDC is possible. In this representation, a partial or incomplete factorization that retains the most dominant contributions is desirable as shown in the plot. Several potential applications can be envisioned.

Remarks:

- Dimensionality reduction is accomplished by a partial or incomplete factorization of the form

$$\widehat{X} \approx \sum_{i=1}^{m} X_i = \sum_{i=1}^{m} \underbrace{t_i p_i^T}_{X_i} = TP^T \tag{9.7}$$

- Once the model (9.2) is obtained, the multiblock representation can be analyzed using any of the multiview methods in Chapter 6.
- Direct analysis of the model has the limitation of becoming intractable as the number of measurements/PDCs increases.

Alternatively, a useful representation of this model can be obtained from SVD analysis of each slice, namely

$$\hat{X} \approx \sum_{i=1}^{m} X_i = \sum_{i=1}^{m} \underbrace{\sigma_i u_i v_i^T}_{X_i} = U \Sigma V^T \tag{9.8}$$

where $X_i = \sigma_i u_i v_i^T$. Note that this latter implementation is equivalent to that based on PCA. The extension to the three-dimensional case is possible using tensor analysis.

In this case, each data set is represented as a third-order tensor

$$\widehat{X} \in \mathfrak{R}^{N \times m_1 \times m_2},$$

where N represents the number of samples and m_1, m_2 denote respectively, the dimensions of each slice of the second-order tensor-based decomposition in Figure 9.3. From tensor analysis theory, a three-mode tensor can be unfolded into a matrix in 3 ways, one for each mode (Wu et al. 2019; Lahat et al. 2015). But, there are several advantages to direct tensor analysis.

Given a tensor representation collected by a WAMS in Figure 9.2, a representation of the form depicted in Figure 9.3 can be obtained as discussed by Wu et al. (2019). Within the singular value decomposition framework it is possible to study various aspects of system behavior in an efficient manner.

Examples of potential applications are summarized in Table 9.1.

Complementary approaches to multiblock analysis are discussed later in Sections 9.3 and 9.4.

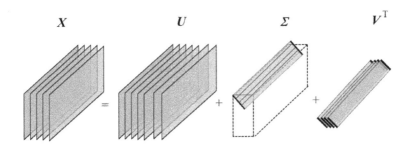

FIGURE 9.3
Tensor-based singular value decomposition. (Adapted from Wu, J. et al., *IEEE Trans. Image Process.*, 28, 5910–5922, 2019.)

TABLE 9.1

Potential Applications of Tensor-Based Decompositions

Application	References
• Missing data estimation	Acar et al. (2011)
• Data mining	Kolda and Sun (2008)
• Analysis of hidden correlations	Sun et al. (2006)
• Improved data fusion	Espadoto et al. (2019)

9.3 Challenges for the Analysis of High-Dimensional Data

High-dimensional power system data analysis poses conceptual and practical challenges that reduce its effectiveness and require careful consideration. A brief discussion of some of these issues might include:

Noise and missing data: High levels of ambient noise and missing data may affect the performance of data mining and data fusion methods. While some dimensionality reduction techniques may suppress noise, data cleansing should be considered as a first step in the application of these techniques.

Similarity measurements: The intrinsic dimension of the input space is much less that the number of collected measurements (Esling and Agon 2012). Metrics are therefore needed to determine redundant measurements as well as to discard relatively insignificant contributions.

Data fusion level: Data fusion techniques can be classified into three main levels: raw data level, feature level, and decision level (Zhang 2010). For power system data determining good observables to monitor and assess power system health is a critical issue that has received limited attention.

Classification and fusion: Data classification is a key area of future research. Classification methods essentially predict group memberships for data instances. Typical approaches include techniques such as decision trees, k nearest neighbors, and support vector machines. A number of algorithms have shown promise in classifying data with high accuracy. Examples include statistical models and sparse representation models (Della Mura et al. 2015).

Extensions to the multimodal case include such techniques as composite kernel SVM, and composite kernel local Fisher's discriminant analysis. Experience with the application of SVMs and other classifiers to complex power system data is deferred till Chapter 10. Adaptive feature selection is another emerging issue in data classification with potential applications to power system data. One example of this need is power system prognosis under cascading events in which data are constantly changing and dynamic patterns may evolve in time. A thorough review of the subject is given by Howard (1989).

Data visualization and visual data mining: Synthesizing vast amounts of data into low-dimensional versions is a challenging problem. A related issue is that of reducing the number of attributes (feature selection) to be analyzed and visualized. By reducing the size of the data to be analyzed and visualized, patterns and correlations are easier to view, and visualization tool performance is optimized. An emerging area within this category involves support vector machines.

Visual data fusion: Another area of research interest is visual data fusion. As noted by Treinish (2000), the human visual system is an integral part of the process of extracting knowledge from complex data. Advantages of these approaches include:

- Effective visualization may prevent artifacts arising from the visualization process to be interpreted as features in the data.

- Further, data utilization by the analyst or operator may be affected by the form in which information is presented.

In the power system literature, many techniques to visualize data have been developed and integrated for power system monitoring.

Adaptive feature selection: As discussed in previous sections, approaches to dimensionality reduction can be broadly characterized as selection-based and projection-based. Techniques to perform selection-based or feature-based are of interest to input-space reduction (Korycinski et al. 2003; Lunga et al. 2014), and variable and feature selection (Guyon and Elisseeff 2003). The motivation for input-space reduction is three-fold: (a) To reduce the data dimensional for data mining and data fusion, (b) to reduce the number of measurements required to characterize patterns in space and time or alternatively to detect irrelevant or redundant variables in the data, and (c) facilitate data visualization and data understanding.

Real-time data fusion: Near real-time monitoring systems have been recently developed and implemented in many power systems based on synchrophasor technology. However, the incorporation of data mining and data fusion techniques to wide-area monitoring architecture presents unique challenges that need to be considered and may affect communication bandwidth and latency. Data fusion, in particular becomes more challenging when sensor networks are deployed in real-time automated systems due to the nature of distributed data.

Probabilistic Fusion: Sensor measurements are uncertain in nature. Uncertainties in sensor measurements arise from diverse factors such as noise, impreciseness and inconsistencies present in the environment (Abdulhafiz and Khamis, 2013). Probabilistic methods (Tipping et al. 1999) hold promise for improving system characterization. Compared to deterministic approaches, the literature on probabilistic methods is relatively sparse.

Change detection: Two main issues are of interest in the analysis of simultaneously recorded signals: change detection and data correlation. The first issue has been discussed in Chapter 8. The latter issue seeks to determine similar data performance that could lead to further dimensionality reduction. A schematic illustration of this idea is given in Figure 9.4.

Enhanced decision support systems: It is only recently that advanced prediction, prognostics and decision support systems are being developed. A major challenge in the development and deployment of early warning systems is the need for scalable time-sensitive data exchange and processing. This is particularly troublesome in the case of application involving heterogeneous data sources (Poslad et al. 2015).

Application to high-dimensional data: In the literature described hitherto, data mining and data fusion methods have been applied to small and medium-size systems. Such methods are very computationally intensive, especially in the presence of complicated spatiotemporal dependence and may be scalable. Due to the massive volumes of data processed, projection-based methods may not be efficiently applied to complex, high dimensional data. Approaches based on deep learning methods discussed in Sections 9.5 and 9.6 hold promise to tackle some of these issues.

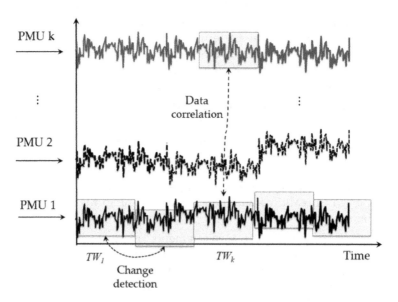

FIGURE 9.4
Illustration of change detection and data correlation.

9.4 Distributed Data Mining

At present, most data mining techniques are essentially centralized. These approaches, however, may be inappropriate or inefficient for many distributed data applications because of the long response times and the inappropriate use of distributed resources (Park and Kargupta 2003; Denham 2019). Data transmission from distributed sensors or microgrids to a central processing unit may create heavy traffic over the bandwidth of communication systems and also increase latency. With the advent of distributed renewable generation and more complex system configurations, there has been a surge of interest in utilizing distributed data mining and data fusion architectures for monitoring system behavior.

Modern distributed architectures and algorithms offer the possibility to incorporate more complex information to global system behavior that needs to be processed taking into account that data may be inherently distributed geographically across the power system.

A conceptual illustration of a distributed data mining architecture is shown in Figure 9.4. Typically, in this architecture, data mining is performed at a local level resulting in two or more local models that need to be aggregated. Several issues make the analysis of distributed data difficult:

- The existence of various constraints such as limited bandwidth, geographical separation, and data privacy (proprietary data) to mention a few issues (da Silva et al. 2005).

- Geographical data may exhibit local trends and be inherently heterogeneous.

- Further, local data may be unrelated to global system behavior and therefore techniques are needed for feature selection prior to data mining and data fusion.

These problems can be efficiently addressed using distributed data mining techniques. As discussed in (Denham et al. 2019), however, this requires the development of specialized algorithms for true distributed data mining. In this regard, SVM and other classifiers can be employed for feature section and feature extraction and data classification prior to the application of data mining techniques. Other associated problems include distributed clustering and distributed data fusion.

Application of distributed data mining architectures has recently been investigated using multiblock analysis techniques. Other approaches based on tensor representations hold promise for the analysis of more complex data sets as discussed in the Sections 9.3 and 9.4.

9.5 Dimensionality Reduction

One important application of data fusion techniques is dimensionality reduction (Engel et al. 2011; Li et al. 2012). Major research issues are briefly summarized in Subsections 9.5.1 and 9.5.2.

9.5.1 Limitations of Existing Dimensionality Reduction Methods

Nonlinear dimensionality reduction methods discussed in Chapter 6 are constructed based on a specific time window. While these approaches can be easily modified to capture local behavior, they are based on matrix representations which make them unsuitable for the study of high-dimensional data. In addition, these methods are prone to higher computational cost.

In this regard, the use of sparse methods for data fusion with the ability to learn from the data is especially interesting. Methods include LLE and other sparse representations (Van der Maaten 2009).

9.5.2 Deep Learning Multidimensional Projections

The application of dimensionality reduction methods to large, high-dimensional data sets is computationally challenging and suffers from various problems. Recent alternatives based on deep learning projections have the capability to handle out-of-sample data in an efficient manner and can be used to learn any projection technique (Espadoto et al. 2019, Gisbrecht et al. 2012).

Figure 9.5 illustrates schematically the nature of this approach.

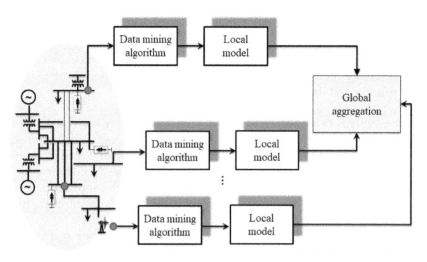

FIGURE 9.5
Distributed data mining architecture. (Based on Park, B., Kargupta, H., Distributed data mining: Algorithms, systems, and applications, in Ye, N. (ed.), *The Handbook of Data Mining*, pp. 341– 358, Lawrence Erlbaum Associates, Mahwah, NJ, 2003.)

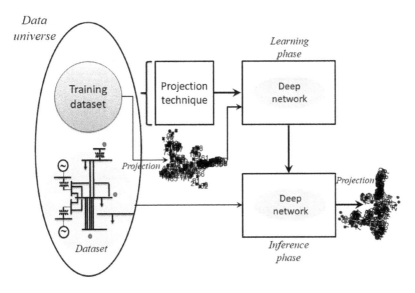

FIGURE 9.6
A graphical depiction of a deep learning projection model. (Adapted Espadoto, M. et al., arXiv.1902.07958, February 2019.)

A property shared by these methods is that they allow learning from the data itself (Worden et al. 2011) and can be more efficient that their counterpart using spectral projection methods which is illustrated in Figure 9.6.

Other contributions include methods to derive low-dimensional models for data visualization based on deep neural networks (Becker et al. 2017) that extend previous work based on the application of neural networks to reduce data dimensionality (Hinton and Salakhutdinov 2006).

9.6 Bio-Inspired Data Mining and Data Fusion

Many recent computational algorithms for data mining and data fusion have been motivated by biological processes. These include neural-network based approaches, evolutionary approaches, artificial intelligence and machine learning techniques (Worden et al. 2011). Interest in this subject is reflected in recent special publications (Olario and Zomaya 2006).

Unlike more conventional methods, deep-learning techniques such as convolutional neural networks, deep belief networks, and recurrent neural networks can learn from the data itself. Therefore these deep-learning techniques can, in some situations, outperform more traditional techniques.

Table 9.2 summarizes some analytical models used in recent applications. Each category may include several subcategories.

TABLE 9.2

Overview of Bio-Inspired Data Mining and Data Fusion Techniques

Model	Applications
PCA/POD	Data fusion
Diffusion maps	Spatiotemporal clustering
	Bio-inspired clustering application Nonlinear PCA
	Data visualization
	Decision support algorithms
	Information visualization
Artificial	Anomaly detection and prediction
intelligence and	Health monitoring systems
machine learning	Predictive and real time analytics
	Data mining
	Situation and threat assessment
Deep learning	Anomaly and change detection
	Data fusion
	Classification
Fuzzy Kalman filter	Multisensory data fusion architectures, clustering and classification

9.7 Other Emerging Issues

Other emerging issues include data fusion under imprecise or unknown environments (Fouratti, 2015), predictive learning, visual analytics, and data fusion via intrinsic dynamic variables (Williams 2015), and anomaly detection using data mining techniques (Agrawal and Agrawal 2015).

Envisaged applications include:

- Clustering based anomaly detection,
- Classification-based anomaly detection, and
- Time scale separation in dimensionality reduction.

9.8 Application to Power System Data

One of the major application areas for data mining and data fusion techniques is power system monitoring. In Fusco et al. (2017), a computational framework is proposed for power systems data fusion—based on probabilistic graphical models and capable of combining heterogeneous data sources with classical state estimations. Arvizu and Messina (2016) explored the use of diffusion maps to characterize the collective dynamics of transient processes in power systems.

Efforts have also been made to develop techniques to jointly analyze multi-type, multisource data. The joint use of frequency and voltage signals has already been investigated for diverse applications. Work such as that shown in Dutta and Overbye (2014) is seminal in this context.

9.8.1 Wind and Renewable Energy Generation Forecasting

With the fast development of the highly variable and uncertain field of renewables generation, accurate forecasting is growing in importance. Figure 9.7 shows one-week of wind generation at a 115 kV point of common coupling of a large wind farm. Developing forecasting techniques for such measurements poses major conceptual and practical challenges due to the natural uncertainty and variability of the wind itself.

Within the last decade several attempts have been made to develop forecasting techniques for renewables generation. In Mohan et al. (2018), a data-driven strategy for short-term electric load forecasting using DMD was proposed. Recently, Zavala and Messina et al. (2014) and Messina et al. (2017) discuss the application of dynamic harmonic regression to predict wind generation in large-scale power systems (Figure 9.8).

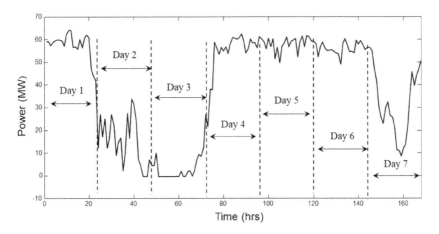

FIGURE 9.7
Weekly data for a 115 kV wind farm.

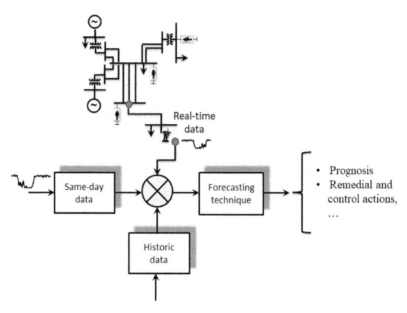

FIGURE 9.8
Illustration of a forecasting technique.

9.8.2 Application to Distribution Systems

Another significant area of continuing research is the application of data mining and data fusion techniques to distribution systems. Because of the increased availability of smart sensors and distribution PMUs (DPMUs), the development of fusion techniques is of great interest (Donde and Mohamed 2016).

Representative applications in power system modal analysis appear in several recent publications (Joseph and Jasmin 2017; Pinte et al. 2015; Meier et al. 2017; Roberts et al. 2016; Donde and Mohamed 2016; Zhang et al. 2018; Wang et al. 2016; Guikema et al. 2010).

Other recent applications include:

- Identification of phase connectivity using data mining and the Open Distribution System Simulator,
- Power quality monitoring,
- Distribution system monitoring,
- Characterization of distributed generation, and
- Phasor-based control, among other issues.

Other emerging applications include monitoring, protection, and control of distribution networks and estimation of power outage risks.

References

Abdulhafiz, A. A., Khamis, A., Handling data uncertainty and inconsistency using multisensory data fusion, *Advances in Artificial Intelligence*, Article ID 241260, 2013.

Acar, E., Kolda, T. G., Dunlavy, D. M., All-at-once optimization for coupled matrix and tensor factorizations, ArXiv.org, 1105.3422, e-print, May 2011.

Agrawal, S., Agrawal, J., Survey on anomaly detection using data mining techniques. *Procedia Computer Science*, 60, 708–713, 2015.

Arvizu, C. M. C., Messina, A. R., Dimensionality reduction in transient simulations: A diffusion maps approach, *IEEE Transactions on Power Delivery*, 31(5), 2379–2389, 2016.

Atluri, G., Karpatne, A., Kumar, V., Spatio-temporal data mining: A survey of problems and methods, arXiv:1711.04710, 2017.

Becker, M., Lippel, J., Stuhlsatz, A., Regularized nonlinear discriminant analysis: An approach to robust dimensionality reduction for data visualization, *12th International Joint Conference on Computer Vision, Imaging and Computer Graphics Theory and Applications*, 3, Porto, Portugal, 2017.

Bisantz, A. M., Finger, R., Seong, Y., Llinas, J., Human performance and data fusion based decision aids, *Proceedings of the 2nd International Conference on Information Fusion based Decision Aids – Fusion'99*. International Society of Information Fusion, Chicago, IL, 1999.

Chen, X. C., Faghmous, J. H., Khandelwal, A., Kumar, V., Clustering dynamic spatio-temporal patterns in the presence of noise and missing data, *Proceedings of the Twenty-fourth International Conference on Artificial Intelligence (IJCAI 2015)*, Buenos Aires, Argentina, pp. 2575–2581, 2015.

Da Silva, J. C., Giannella, C., Bhargava, R., Kargupta, H., Klusch, M., Distributed data mining and agents, *Engineering Applications of Artificial Intelligence*, 18, 791–805, 2005.

Della Mura, M. D., Prasad, S., Pacifici, F., Gamba, P., Benediktsson, J. A., Challenges and opportunities of multimodality and data fusion in remote sensing, *Proceedings of the IEEE*, 103(9), 11(4), 1585–1601, 2015.

Denham, B., Pears, R., Asif Naee, M., HDSM: A distributed data mining approach to classifying vertically distributed data streams, *Knowledge-Based Systems*, 189, 105114, 2019.

Donde, V., Mohamed, S., Data fusion and analytics applications for PG&E's power distribution systems, i-PCGRID workshop, https://ipcgrid.ece.msstate.edu/presentations/2016/, Mississippi State University, 2016.

Dutta, S., Overbye, T., Feature extraction and visualization of power system transient stability results, *IEEE Transactions on Power Systems*, 29(2), 966–973, 2014.

Engel, D., Huttenberger, L., Hamann, B., A survey of dimension reduction methods for high-dimensional data analysis and visualization, *Visualization of Large and Unstructured Data Sets: Applications in Geospatial Planning, Modeling and Engineering–Proceedings of IRTG* 1131, Dagstuhl Publishing, Germany, pp. 135–149, Workshop 2011.

Esling, P., Agon, C., Time-series data mining, *ACM Computing Surveys, Association for Computing Machinery*, 45(1), A:1–A:31, 2012.

Espadoto, M, Hirata,1 N. S. T., Telea, A. C., Deep learning multidimensional projections, arXiv.1902.07958, February 2019.

Fouratti, H. (Ed.), *Multisensor Data Fusion – From Algorithms and Architectural Design to Applications*, CRC Press, Boca Raton, FL, 2015.

Fusco, F., Tirupathi, S., Gormally, R., Power systems data fusion based on belief propagation, *2017 IEEE PES Innovative Smart Grid Technologies Conference Europe (ISGT-Europe)*, Torino, Italy, September 2017.

Ghamisi, P., Rasti, B., Yoyoka, N., Wang, Q., Hofle, B., Bruzzone, L., Bovolo, F. et al., Multisource and multitemporal data fusion in remote sensing, *IEEE Geoscience and Remote Science Magazine, IEEE Geoscience and Remote Science Magazine, IEEE Geoscience and Remote Sensing Magazine*, pp. 6–39, March 2019.

Gisbrecht, A., Lueks, W., Mokbel, B., Hammer, B., Out-of-sample extensions for nonparametric dimensionality reduction, *ESANN 2012 Proceedings, European Symposium on Artificial Neural Networks*, Computational Intelligence and Machine Learning, Bruges, Belgium, April 2012.

Guikema, S. D., Quiring, S. M., Han, S. R., Prestorm estimation of hurricane damage to electric power distribution systems, *Risk Analysis*, 30(12), 1744–1752, 2010.

Guyon, I., Elisseeff, A., An introduction to variable and feature selection, *Journal of Machine Learning Research*, 3, 1157–1182, 2003.

Hinton, G. E., Salakhutdinov, R. R., Reducing the dimensionality of data with neural networks. *Science*, 313, 504–507, 2006.

Joseph, S., Jasmin, E. A., Big data analytics for distribution system monitoring in smart grid, *International Journal of Smart Home*, 11(5), 21–32, 2017.

Kisilevich, S., Mansmann, F., Nanni, M., Rinzivillo, S., Spatio-temporal clustering: A survey, in Maimon, O., Rokach, L. (Eds.), *Data Mining and Knowledge Discovery Handbook*, Springer Science, New York, 2010.

Kolda, T. G., Bader, B. W., Tensor decompositions and their applications, *SIAM Review*, 51(3), 455–500, 2009.

Kolda, T. G., Sun, J. Scalable tensor decompositions for multi-aspect data mining, *2008 Eighth IEEE International Conference on Data Mining*, Pisa, Italy, December 2008.

Korycinski, D., Crawford, M. M., Barnes, J. W., Adaptive feature selection for hyperspectral data analysis, *IEEE International Geoscience and Remote Sensing Symposium*, Toulouse, France, July 2003.

Lahat, D., Adali, T., Jutten, C., Multimodal data fusion: An overview of methods, challenges and prospects, *Proceedings of the IEEE*, 103(9), 1449–1477, September 2015.

Li, W., Prasad, S., Fowler, J. E., Bruce, L. M., Locality-preserving dimensionality reduction and classification for hyperspectral image analysis, *IEEE Transactions on Geoscience and Remote Sensing*, 50(4), 1185–1198, 2012.

Lunga, D., Prasad, S., Crawford, M. M., Ersoy, O., Manifold-learning based feature extraction for classification of hyperspectral data, *IEEE Signal Processing Magazine*, 31(1), 55–66, 2014.

Meier, A., V, Stewart E., McEachern, A., Andersen, M., McEachern, L., Precision Micro-Synchrophasors for Distribution Systems: A Summary of Applications, *IEEE Transactions on Smart Grid*, 8(6), 2926-2936, November 2017.

Messina, A. R., Castellanos, R., Castro, C. M., Barocio, E., Jiménez Zavala, A., Large-scale wind generation development in the Mexican power grid: Impact studies, in *Handbook of Distributed Generation*, Vol. 13, pp. 109–148, Springer International Publishing, Cham, Switzerland, 2017.

Messina, A. R., *Wide-Area Monitoring of Interconnected Power Systems*, IET Power and Energy Series 77, London, UK, 2015.

Mohan, N., Soman, K. P., Kumar, S. S., A data-driven strategy for short-term electric load forecasting using dynamic mode decomposition model, *Applied Energy*, 232, 229–244, 2018.

Olario, S., Zomaya, A. Y. (Eds.), *Handbook of Bioinspired Algorithms and Applications*, Chapman Hall & Hall/CRC Computer and Information Science Series, Boca Raton, FL, 2006.

Papalexakis, E. E., Faloutsos, C., Sidiropoulos, N. D., Tensors for Data Mining and Data Fusion: Models, Applications, and Scalable Algorithms. *ACM Transactions on Intelligent Systems and Technology*, 8(2), 16:1–16:44, 2016.

Park, B., Kargupta, H. Distributed Data Mining: Algorithms, Systems, and Applications, In Ye, N. (ed.) *The Handbook of Data Mining*, pp. 341–358, Lawrence Erlbaum Associates, Mahwah, NJ, 2003.

Pinte, B., Quinlan, M., Reinhard, K., Low voltage micro-phasor measurement unit (µPMU), *2015 IEEE Power and Energy Conference at Illinois (PECI)*, Champaign, IL, February 2015.

Poslad, S., Middleton, S. E., Chaves, F., Tao, R. Necmioglu, O., Bugel, A. R., A semantic IoT early warning system for natural environment crisis management, *IEEE Transactions on Emerging Topics in Computing*, 3(2), 246–257, 2015.

Roberts, C. M., Shand, C. M., Brady, K. W., Stewart, E. M., McMorran, A. W., Taylor, G. A., Improving distribution network model accuracy using impedance estimation from micro-synchrophasor data, *2016 IEEE Power Engineering Society General Meeting*, Boston, MA, July 2016.

Shekhar, S., Jiang, Z., Ali, R. Y., Eftelioglu, E., Tang, X., Gunturi, V. M. V., Zhou, X., Spatiotemporal data mining: A computational perspective, *ISPRS International Journal of Geo-Information*, 4, 2306–2338, 2015.

Sun, J., Tao, D., Faloutsos, C., Beyond streams and graphs: Dynamic tensor analysis, *Proceedings of the Twelfth ACM SIGKDD International Conference on Knowledge Discovery and Data Mining*, Philadelphia, PA, August 20–23, 2006.

Taniar, D., *Data Mining and Knowledge Discovery Technologies*, IGI Publishing, Hershey, PA, 2008.

Tipping, M. E., Bishop, C. M., Probabilistic principal component analysis, *Journal of the Royal Statistical Society B*, 61(3), 611–622, 1999.

Treinish, L. A., Visual data fusion for decision support applications of numerical weather prediction, *Proceedings of the Conference on Visualization'00*, pp. 477–480, Los Alamitos, CA, 2000.

Van der Maaten, L. J. P., Postma, E. O., Ven den Herik, H. J., Dimensionality reduction: A comparative review, Tilburg University Technical Report, Holland, TiCC-TR 2009-005, 2009.

von Meier, A., Stewart E., McEachern, A., Andersen, M., McEachern, L., Precision micro-synchrophasors for distribution systems: A summary of applications, *IEEE Transactions on Smart Grid*, 8(6), 2926–2936, 2017.

Wang, W., Yu, N., Foggo, B., Davis, J., Phase identification in electric power distribution systems by clustering of smart meter data, *2016 15th IEEE International Conference on Machine Learning and Applications (ICMLA)*, Anaheim, CA, December 2016.

Williams, M. O., Rowley, C. W., Mezicm I., Kevrekidis, I. G., Data fusion via intrinsic dynamic variables: A application of data-driven Koopman spectral analysis, *Europhysics Letters (EPL)*, 109(4), 40007p1–p6, 2015.

Worden, K., Staszewski, W. J., Hensman, J. J., Neural computing for mechanical systems research: A tutorial overview, *Mechanical Systems and Signal Process*, 25, 4–111, 2011.

Wu, J., Lin, Z., Zha, H., Essential tensor learning for multi-view spectral clustering, *IEEE Transactions on Image Processing*, 28(12), 5910–5922, December 2019.

Zavala, A. J., Messina. A. R., A dynamic harmonic regression approach to power system modal identification and prediction, *Electric Power Components and Systems*, 42(13), 1474–1483, 2014.

Zhang, J., Multi-source remote sensing data fusion: Status and trends, *International Journal of Image and Data Fusion*, 1(1), 5–24, 2010.

Zhang, Y., Huang, T., Bompard, E. F., Big data analytics in smart grids: A review, *Energy Informatics*, 1, 1–8, 2018.

10

Experience with the Application of Data Fusion and Data Mining for Power System Health Monitoring

10.1 Introduction

Characterization of the dynamic phenomena that arise when the system is subjected to a perturbation is important in real-time power system monitoring and analysis. In this chapter, simulated data from a large-scale test power system are used to investigate the applicability of data fusion and data mining techniques to the analysis of very large data sets and monitoring of power system dynamic behavior. The challenges faced during the practical implementation of data mining and data fusion techniques are examined and a physical explanation of some algorithms is provided.

Several algorithms are compared in terms of their effectiveness in addressing problems and achieving solutions. Numerical issues are also discussed. Discussions are supported by detailed simulation results to highlight the main issues in the practical implementation of the methods.

10.2 Background

The practical application of data mining and data fusion techniques to large data sets is tested on transient stability simulations of a large test power system. Figure 10.1 is a simplified single-line diagram of the study system showing the main areas of concern for this study. Key buses and major transmission paths are indicated on the schematic map.

The test system includes the detailed representation of 635 generators, 26 large wind farms, 5449 buses, and 5292 transmission lines—see Messina et al. (2006) for detailed discussions of this test system.

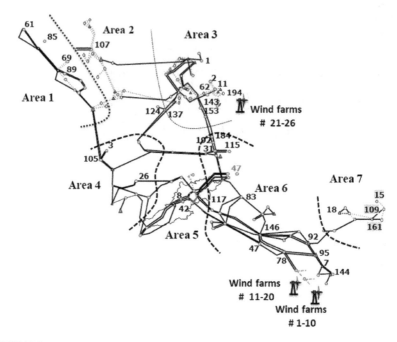

FIGURE 10.1

Simplified layout of the study system showing major transmission resources. Dashed black lines represent approximate regional boundaries.

10.2.1 Wide-Area Measurement Scenarios

The study uses data from 7 regional systems to evaluate the application of data mining and data fusion methods for power system monitoring. Data sets of key system variables from step-by-step simulations are used to assess the ability of data mining and data fusion techniques to monitor system behavior. Selected measurements of interest include:

- Active and reactive power measurements at synchronous machines and other dynamic devices and transmission elements,
- Bus voltage magnitudes and phases,
- Generator and bus frequency deviations, and
- Active and reactive load variations.

Table 10.1 synthesizes the measurements used in the simulations.

Data mining and data fusion techniques are going to be tested on selected combinations of this data to extract dynamic patterns and monitor system behavior.

TABLE 10.1

Wide-Area Measurements Selected for Study

Measurement Type	Dimension	Data Matrix Description
Bus voltage magnitudes at transmission buses	194	$X_{bus\_volt} \in \mathfrak{R}^{3603 \times 194}$
Speed deviations at major generators	138	$X_{speed\_gen} \in \mathfrak{R}^{3603 \times 138}$
Speed deviations at wind farms	26	$X_{speed\_wfs} \in \mathfrak{R}^{3603 \times 26}$
Bus frequencies	174	$X_f \in \mathfrak{R}^{3603 \times 174}$
Wind farms active power	26	$X_{p\_wfs} \in \mathfrak{R}^{3603 \times 26}$

10.2.2 Basic Modal Properties

Two different approaches to global modal damping estimation are employed to serve as reference to evaluate the accuracy and applicability of mining and fusion techniques: (1) Small signal analysis of the linearized power system model, and (2) Koopman/DMD analysis of selected contingency scenarios. These include:

Scenario 1: A base case with the system intact,

Scenario 2: Loss of generator 23 in Area 6,

Scenario 3: Loss of load at bus 153,

Scenario 4: Three-phase fault at bus 153,

Scenario 5: Double-line outages.

The results of small signal calculations for the test system (Scenario 1) are presented in Table 10.2. Slow system dynamic behavior of the test system is characterized by four principal inter-area modes at 0.39, 0.54, 0.69, and 0.73 Hz. The slowest mode represents an oscillation in which machines in Areas 6

TABLE 10.2

The Slowest Modes of Oscillation of the Test System
Base case

Eigenvalue	Frequency (Hz)	Damping in %
1	0.394	0.357
2	0.545	7.001
3	0.696	7.46
4	0.730	1.583
5	0.748	1.522

TABLE 10.3

DMD Analysis for Scenario 3

Mode Number, j	Frequency	Damping	Energy (E_j)
1	0.394	0.357	
2	0.545	7.001	
3	0.696	7.46	
4	0.730	1.583	
5	0.748	1.522	

and 7 swing in opposition in Areas 1 through 3—refer to Messina et al. (2006) for a detailed discussion of the nature of these modes.

For reference and validation, Table 10.3 shows the extracted modes using Koopman/DMD analysis for scenario 3 above. Similar results are obtained for other operating conditions. Figure 10.2 shows the speed-based mode shapes for the slowest inter-area mode at 0.394 Hz in Table 10.3 extracted using Koopman analysis.

As shown in Barocio et al., (2015), within the framework of DMD analysis, the observed system behavior is expressed as

$$x_k = x(t_k) = Re\left\{\sum_{j=1}^{m} \lambda_j \phi_j a_j(t)\right\} = \sum_{j=1}^{m} a_j \phi\left[e^{(\sigma_j + i\omega_j)} + e^{(\sigma_j - i\omega_j)}\right] \qquad (10.1)$$

$k = 1, 2, \ldots, m$, where j and m denote the number of measurements, and σ_j and ω_j are the modal damping and frequency computed from

$$\sigma_j = Re\left\{\log(\lambda_j)\right\} / \Delta t$$
$$\omega_j = Im\left\{\log(\lambda_j)\right\} / \Delta t / 2\pi \qquad (10.2)$$

A ranking criterion based on the energy-based metric

$$E_j = \frac{1}{T}\|\phi_j\|^2 \int_0^T \lambda_j^{2\sigma_j t} dt = \|\phi_j\|^2 \frac{e^{2\sigma_j T} - 1}{2\sigma_j T} \qquad (10.3)$$

is adopted to rank variables (states) and construct energy-based clusters—refer to Chapter 4 for details.

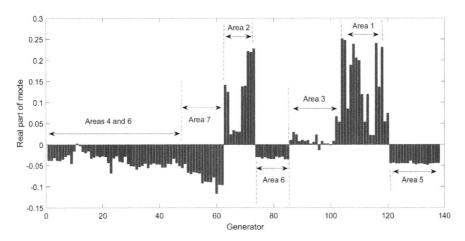

FIGURE 10.2
Plot of the speed components of the eigenvector of inter-area mode 1. Data set $X_{speed\_gen} \in \mathfrak{R}^{3603 \times 138}$.

Results are found to be in good agreement with the modal results in Table 10.2.
Having computed the modal parameters, the measured data can be reconstructed using the truncated approximation

$$X(t) = [x_1 \quad x_1 \dots \quad x_m] = \sum_{j=1}^{m} \lambda_j \phi_j a_j(t) \tag{10.4}$$

In the present context, since the time series contain 3603 observations, application of the DMD algorithm from Chapter 4 for the data set $X_g = [X_{speed\_gen}] \in \mathfrak{R}^{3603 \times 138}$ results in 138 modes for the case of speed deviations in Table 10.1, while Koopman analysis yields 3603 modes.

For the purpose of this study, attention is focused on the two slowest modes of the system at 0.394 and 0.545 Hz which capture the highest energy; the top five most energetic modes are then selected for clustering. Both, actual PMU locations and simulated voltage signals that are typical of other buses in terms of modal content are selected.

10.3 Sensor Placement

As emphasized in previous Chapter 6, sensor placement plays a critical role in the ability of sensors to monitor system behavior. Drawing upon previous research, transient stability simulations and actual PMU locations for the test systems, 194 bus voltage signals spanning a 30-second period, representing the post-disturbance condition, were selected for sensor placement for several realistic system disturbances and loading condition scenarios. Both, transmission buses and wind farms terminal buses are selected for analysis.

From Table 10.1, the snapshot (measurement) data is defined as

$$X_V = \begin{bmatrix} V_1 & V_2 \dots & V_{194} \end{bmatrix}^T = \begin{bmatrix} X_{bus_{volt}} & X_{bu\_volt\_wfs} \end{bmatrix}^T, \tag{10.5}$$

with $V_j = \begin{bmatrix} V_j(t_0) & V_j(t_1) \dots & V_j(t_N) \end{bmatrix}$, and $N = 3603$.

The objective is to determine sensor locations that result in accurate characterization of dominant inter-area modes within a specified frequency band (0.1–1.0 Hz), and the reconstruction of selected states using the selected sets of measurements.

The placement problem is formulated as the solution of the optimization problem

$$\varepsilon'_{min} = \frac{min}{\{k\} \in \{x\}} \sum_{i=1}^{M} \varepsilon_{c\_i}(\{k\})$$

where the sum of the error ε_{c_i} for M modesis the objective function, and the set $\{x_1, x_2, \dots, x_m\}$ denotes all elements of observation or measurement points.

Following Alonso et al. (2004) and Messina (2015), the goal is to determine a low-dimensional subspace that captures most of the relevant dynamics of the original physical space.

Table 10.4 shows the top fifteen candidates for sensor locations determined using the approach just discussed for various contingency scenarios ranging from three-phase faults at major buses to load shedding. In all cases, the energy captured by the dominant mode is over 60% for the scenarios in Table 10.1.

TABLE 10.4

Ranking of Candidate Sensor Locations for Various Disturbance Scenarios. Data set X_V in (10.5)

Contingency Scenario	Ranking of Candidate Sensor Locations
Scenario 1	144, 135, 66, 107, 100, 78, 65, 105, 23, 172, 47, 191, 175, 176, 156
Scenario 2	144, 135, 107, 66, 94, 36, 100, 156, 65, 105, 23, 78, 21, 145, 75
Scenario 3	144, 135, 66, 107, 100, 78, 65, 105, 23, 75, 156, 47, 146, 22, 74
Scenario 4	144, 135, 107, 66, 100, 65, 23, 105, 78, 75, 22, 156, 90, 74, 145

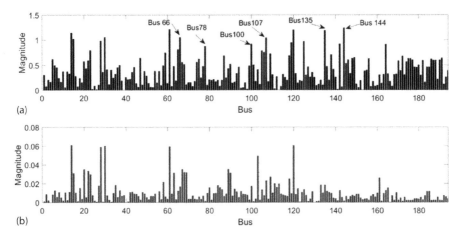

FIGURE 10.3
Magnitudes of voltage-based eigenvector shapes computed using Koopman analysis. (a) Inter-area mode 1, and (b) Inter-area mode 2.

Results are found to be consistent and identify buses 144, 135, 66, 100, and 107 as the best options to place sensors.

As it was suggested in Chapter 3, voltage-based eigenvectors provide a first useful estimate of suitable locations to place sensors. In addition, modal voltage deviations provide complementary information to that available from the sensor placement information.

To test this concept, pseudo voltage mode shapes were computed from the data set X_V using Koopman/DMD analysis.

Figure 10.3 shows the voltage eigenvectors of the slowest inter-area modes in Table 4.1, computed using DMD. Similar results are obtained for the other scenarios in Table 10.4. For easiness of visualization, only the magnitudes are presented.

Simulation studies in Figure 10.3 show that buses 144, 135, 107, and 66 have a large magnitude. For inter-area mode 2, the analysis singles out buses 61, 120, and 16. Note that the candidate set of ranking locations in Figure 10.3a and b includes bus voltage swings associated with inter-area modes 1 and 2 as expected—refer to Table 10.4.

In Sections 10.4 and 10.5, simulated data from major disturbances are used to investigate the ability of data mining and data fusion techniques to monitor and visualize system behavior. Several multisensor mining and fusion techniques are investigated and tested.

10.4 Cluster-Based Visualization of Transient Performance

The multivariate clustering methods outlined in Chapter 3 were used to identify critical areas for system monitoring. Two modified data scenarios are initially considered for application of these techniques:

> *Case 1.* A simplified case. The data used for this study consists of selected generator rotor frequency measurements from 40 major generators and 26 wind farms. This case could be considered a reduced order model of the detailed case described below.
>
> *Case 2.* A detailed data case, in which the speed deviation of 136 generators (data vector $X_{\text{speed\_gen}}$) and 26 wind farms ($X_{\text{speed\_wfs}}$) are used to simultaneously determine generator clusters.

In both cases, a spatiotemporal representation was adopted, and the clusters were determined using different approaches described in previous chapters. Figure 10.4 shows the time histories of speed deviations of the generators and wind farms. Examination of the data in Figure 10.4b and c suggests that a model of the form $m(t) \approx \sum_{j=1}^{r} C_j \cos(\omega_j t + \varphi_j)$, with $r \approx 1,2$ may be appropriate to represent the local signals' trends.

10.4.1 Case 1

A visualization of the clusters extracted from the time histories using three representative techniques (Markov chains, DMD, and DMs) is presented in Figures 10.5 through 10.7. Similar results were obtained using other

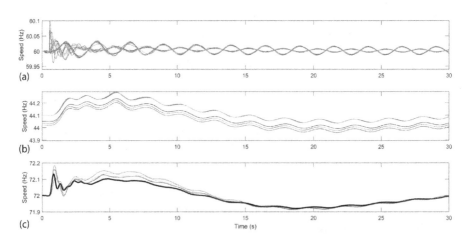

FIGURE 10.4
Wind farm speed deviation for Case 1. (a) Synchronous machines, (b) Wind farms # 1–20, and (c) Wind farms # 21–26).

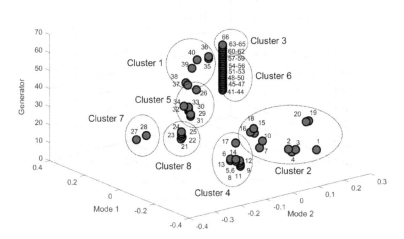

FIGURE 10.5
Cluster representation from DMD analysis. Markov model, Case 1.

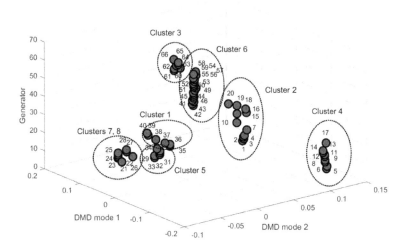

FIGURE 10.6
Cluster representation from DMD analysis. DMD model, Case 1.

techniques and, therefore, that data is not included. In this analysis, the concatenated measurement data is defined as

$$X_g = \begin{bmatrix} X_{\text{speed\_gen}} & X_{\text{speed\_wfs}} \end{bmatrix} \in \Re^{3603 \times 164} \tag{10.6}$$

Numerical experience suggests that the inclusion of wind farms active power may improve the quality of the estimates, but this is not shown in the analysis.

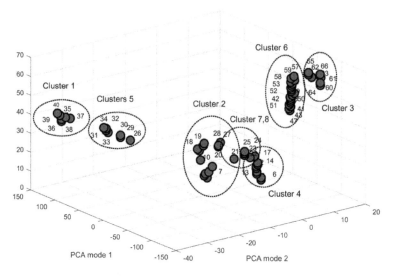

FIGURE 10.7
Cluster representation from DMD analysis. Diffusion maps. Case 1.

In all cases, spatiotemporal clustering of trajectory data above results in 8 coherent groups. Results are only representative since the performance of the methods is dependent on various tuning parameters (e.g., the bandwidth parameter ε for DMs and other spectral analysis techniques, the inflation parameter for Markov chains, etc.).

Once the critical variables are identified, prediction tools can be efficiently applied to the reduced-order model as discussed in Chapter 7.

Diffusion maps are especially well suited to detect good global reduced coordinates and are used here to characterize dynamic phenomena that occur on two or more time scales. From the developed theory in Chapter 3, the data set $\boldsymbol{X}_g$ can be approximated as

$$\widehat{\boldsymbol{X}}_g \approx \boldsymbol{X}_g = \boldsymbol{a}_o(t)\boldsymbol{\psi}_o^T + \sum_{k=1}^{d} \boldsymbol{a}_k(t)\boldsymbol{\psi}_k^T \qquad (10.7)$$

where $\widehat{\boldsymbol{X}}_g$ is an estimate of $\boldsymbol{X}_g$, and the other parameters in Equation (10.7) have the usual interpretation.

Figure 10.8 shows the first four time-dependent amplitude coefficients, extracted using the procedure shown in Arvizu and Messina (2016). Careful examination of the temporal coefficients in Figure 10.8 shows four dominant oscillatory patterns: a slow pattern associated with the slowest mode (Figure 10.8a) at about 0.394 Hz, a slow mode at about 0.682 Hz in Figure 10.8b, and a short-lived mode at about 0.71 Hz, and a trend related to the time

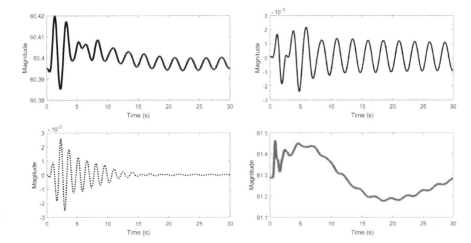

FIGURE 10.8
Temporal coefficients in (10.5). Case 1.

TABLE 10.5

Multisignal Prony Analysis Fit of Temporal
Coefficients in Figure 10.8

Eigenvalue	Frequency	Damping
1	0.394	0.416
2	0.553	8.30
3	0.682	7.53
4	0.710	7.82

evolution of wind farms (Figure 10.8d). Other components are found to be negligible with respect to the dominant modes.

Further, Table 10.5 shows the multisignal Prony results for the data matrix $X = \left[(a_o)^T \quad (a_1)^T \quad (a_2)^T \quad (a_3)^T \right]$. Results are in good agreement with the unreduced system results in Table 10.2 showing that the model accurately captures the temporal behavior of the four slowest inter-area modes.

For completeness and to enable comparison with previous results, the c-means algorithm was also applied to the data set, X_g. Table 10.6 lists the clusters extracted using this algorithm while Figure 10.9 depicts the time evolution of the centroids.

The results lead to the following conclusions:

- Fuzzy c-means clustering allows the identification of two major coherent groups: The group of machines in Areas 1, 2, and 3, and the group of machines in the south systems (Areas 4 through 7).

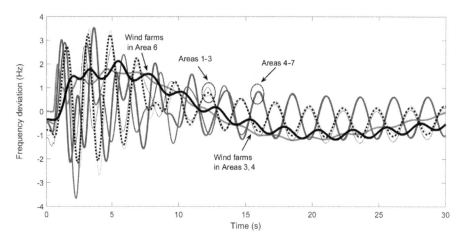

FIGURE 10.9
Time evolution of the *c*-means cluster centers ($m = 8$). Case 1.

TABLE 10.6

Results of Clustering Speed Trajectory Data Using Fuzzy
c-Means Clustering Algorithm. Case 1 above

Cluster	Machines	Geographical Location
1	35–40	Area 6
2	1, 2, 3, 4, 7, 10, 15, 16, 18	Area 6
3	19, 20	Area 6
4	60–66	Area 2
5	5, 6, 8–14, 17	Area 6
6	26, 29, 30–34	Areas 2 and 3
7	41–59	Area 6
8	27, 28	Area 6

Using this approach, the relative contribution (and phase relationship) of the major generators and the wind farms can be
determined.

- By selectively choosing the number of clusters *m*, several reduced-
  order representations can be selected to capture specific behavior.
  Comparison of graphical results in Table 10.6, with candidate sensor
  locations from Table 10.4 shows that clustering techniques can be
  used to capture dominant behavior.

- Physically, fuzzy c-means clustering gives a minimal representation framework which is similar to the OMIB-based representations (Juarez et al. 2006). The window of data used for clustering is 20s. When compared with other approaches (i.e., K-means), fuzzy c-means clustering is faster, easier to interpret, and enables further processing of the extracted centroids for further analysis.
- As shown in these results, the c-means clustering approach provides complementary information to that available from the projection-based technique. This method calculates a predefined number of centroids and evaluates the membership of the individual trajectories to the centroids using a suitable metrics.

Such considerations may have important implications for the design of monitoring architectures based on synchrophasor technology. It is apparent from Figures 10.8 and 10.9 that integrating data from synchronous machines and wind farms results in a better estimation and characterization of the inter-area mode phenomenon.

Envisaged applications of this framework include:

- Hierarchical clustering of dynamic trajectories,
- Identification of outliers and anomalous events, and
- Modal analysis of major inter-area modes.

While not discussed in the present analysis, the extracted centroids agree well with the application of multi-OMIB-based approaches (Juarez et al. 2006). Projection-based approaches, on the other hand, provide a more visual representation of system behavior, but they do not allow the study of interactions between geographical areas. Compared to the K-means clustering algorithm, the c-means clustering approach is computationally efficient, particularly for a large number of trajectories. The example also illustrates that although the volume of measured data may be large, use of only a few underlying measurements or variables can capture global motion.

Figure 10.10 depicts the geographical locations associated with the extracted clusters. By varying the number of clusters, a finer interpretation of system dynamics can be obtained.

Prony analysis of the time evolution of the centroid in Table 10.7 shows the validity and accuracy of the proposed framework; as expected from physical considerations, the analysis shows that the simplified 8-cluster model, effectively captures the dynamics of the first and third inter-area modes.

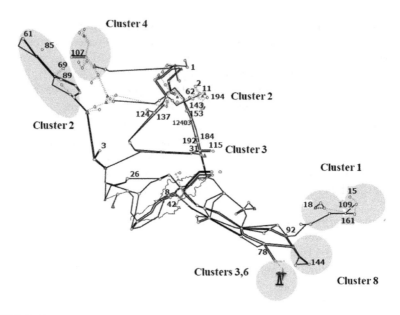

FIGURE 10.10
Spatial distribution of the c-means clusters ($m = 8$). Case 1.

TABLE 10.7

Prony Analysis Fit of Temporal Centroids in Figure 10.9. Case 1 Above

Eigenvalue	Frequency	Damping
1	0.395	0.468
2	0.538	11.208

10.4.2 Case 2

The efficacy of clustering techniques to capture system dynamic behavior for case 2 is illustrated in Figure 10.11 using the t-SNE dimension method. Similar results are obtained using other nonlinear projection methods.

Numerical experience shows, for this case, that nonlinear dimensionality reduction methods give better embedding results. Direct comparisons of these methods, however, is difficult since the performance of these techniques depends on a number of parameters and choices.

For completeness, Table 10.8 shows results of the c-means clustering algorithm for $m=8$. This number of clusters is chosen in order to compare these results with the projection-based techniques in Figures 10.5 through 10.7. The results are summarized in Table 10.8.

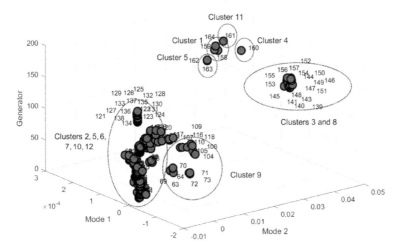

FIGURE 10.11
3D visualization of the speed-power data set obtained using tSNE algorithm.

TABLE 10.8

Clustering of speed-based trajectories for case 2 above using the fuzzy *c*-means clustering algorithm

Cluster	Machines	Geographical Location
1	158, 159, 164	Area 6
2	147–148, 152–153, 155	Area 6
3	144–145, 149–151, 156	Area 2
4	160	Area 2
5	162–163	Area 7
6	1–29, 34–40, 43–46, 49, 74–85, 96, 121–125	WFs, Areas 6 and 3
7	126–138	Area 6
8	27, 28	Area 6

As in the previous case, the centroid-based model can be used to design wide-area monitoring (control) schemes, in which the fundamental system motion is retained. These results may also be relevant to the interpretation of intra-system motion.

Attention is now turned to the problem of multivariate data fusion.

10.5 Multimodal Fusion of Observational Data

Results from the Sections 10.3 and 10.4 emphasize the important of fusing heterogeneous data. In this section, simulated data from the 7 regional systems are utilized to monitor oscillatory behavior. Two multiblock analysis techniques are analyzed but their results are thought to be representative of other approaches: (1) Multiview diffusion maps, and (2) Multiblock PCA.

Figure 10.12 is a conceptual representation of the test system showing the adopted measurement architecture for system monitoring. As discussed in previous chapter, four main layers are incorporated including signal processing, fusion process, and decision support.

There are a number of general points to be made about this model:

- Inputs to the data fusion are high-dimensional data. These data must be consistent in terms of units, dimensions, and magnitude, which requires the use of scaling and other techniques.

- Trends and noise may affect the performance of the techniques.

- Without loss of generality, the adopted approach is particularly suitable to treat multiple signals obtained from a common originating event.

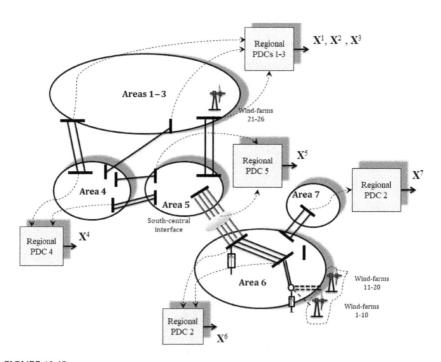

FIGURE 10.12
Conceptual model for the application of multivariate analysis methods.

To be consistent with previous studies, the measurement data for the various regional systems are defined as

$$X_j = \begin{bmatrix} x_1^j(t) \\ \vdots \\ x_{m_j}^j(t) \end{bmatrix} = \begin{bmatrix} x_1^j(t_1) & \cdots & x_1^j(t_N) \\ \vdots & \ddots & \vdots \\ x_{m_j}^j(t_1) & \cdots & x_{m_j}^j(t_N) \end{bmatrix} \in \Re^{N \times m_j}, \; j = 1, \ldots, M \quad (10.8),$$

where m_j represents the number of sensors for area j, and $M = 7$. Note that, in general, $m_i \neq m$, and N is dependent on the measurement sets being considered.

10.5.1 Multiblock PCA Analysis

The purpose of this section is to demonstrate the application of spectral methods to analyze multiple data sets; In the first case, concatenated data is represented in the concatenated form

$$\hat{X} = \begin{bmatrix} X^1 & X^2 \ldots & X^L \end{bmatrix}, \; L = 7. \quad (10.9)$$

Multivariate statistical analysis techniques are then applied to the data to identify dynamic trends of interest.

The analysis technique is general and can be used to characterize fault-dependent mode shapes from detailed transient stability simulations. Because the mode shapes are determined directly from the transient stability output, modeling assumptions are essentially those in the transient stability simulation.

The concatenated data is defined as

$$X_g = \begin{bmatrix} X_{\text{speed\_gen}} & X_{\text{speed\_wfs}} & X_{\text{bus\_volt}} & X_{\text{pot\_wfs}} \end{bmatrix} \in \Re^{3603 \times 384}. \quad (10.10)$$

10.5.2 Multiview Diffusion Maps

Inspired by other multiview integration techniques, the symmetric positive definite multiview kernel can be defined as

$$\hat{K} = \begin{bmatrix} 0_{M \times M} & K^1 K^2 & K^1 K^3 & \cdots & K^1 K^p \\ K^2 K^1 & 0_{M \times M} & K^2 K^3 \cdots & & K^2 K^p \\ K^3 K^1 & K^3 K^2 & 0_{M \times M} \cdots & & K^3 K^p \\ \vdots & \vdots & \ddots & & \vdots \\ K^p K^1 & K^p K^2 & K^p K^3 \cdots & & 0_{M \times M} \end{bmatrix} \in \Re^{l \times m \times l \times m}. \quad (10.11)$$

Like the univariate case, matrix $\widehat{K}$ is symmetric, positive definite, and has a spectral decomposition of interest. Similar to the single type case, the kernels $K^1, \ldots, K^L$ are normalized and then the spectral decomposition can be obtained.

To perform the analysis, the multiview diffusion Gaussian kernel for each data set is defined as

$$\left[K^l \right] = \left[k_{ij}^l \right] = \exp\left(-\frac{\left\| x_i^l(t) - x_j^l(t) \right\|}{\varepsilon_i \varepsilon_j} \right), l = 1, \ldots, L \qquad (10.12)$$

where the $\varepsilon_i \varepsilon_j$ represent local scales associated with data set X^l.

As a first step toward the construction of the model, affinity matrices were computed for the various scenarios of interest. Figure 10.13 shows a heat map of the speed-based pairwise similarity matrix.

Given the pairwise adjacency matrix and the associated matrix, several dimensionality reduction techniques are applied.

Based on the measured speed deviations in Section 10.2.1, MB-PCA was applied to extract modal information from the generator and wind data. For simplicity of exposition, the data was divided into two subblocks: speed deviations at major synchronous machines and speed deviations at the wind farms. The analysis is focused on two main aspects, namely, mode shape estimation and the identification of phase relationships between conventional generators and wind farms.

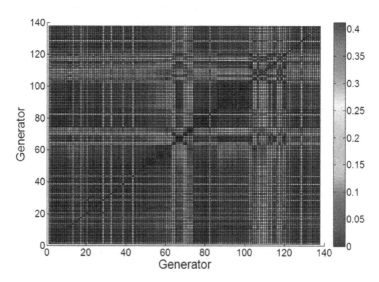

FIGURE 10.13
Visualization of speed-based, pairwise similarity matrix computed using diffusion maps.

In this analysis, the data matrix can be represented as

$$X_g = T_{sup}P_{sup}^T + E \tag{10.13}$$

Shown in Figure 10.8 is a bar representation of the first principal component in the super-score matrix, T_{sup}. The analysis discloses three main oscillation clusters associated with machines in the north and south systems. When compared with the mode-shape extracted using DMD in Figure 10.3, it is apparent that multiblock techniques can accurately capture global system motion.

The weight matrix **W** for this example is:

$$W = \begin{bmatrix} w_1^T \\ w_1^T \end{bmatrix} = \begin{bmatrix} 0.3838 & 0.9087 & 0.9725 & 0.9736 & 0.5302 \\ 0.9234 & 0.4174 & 0.2331 & 0.2282 & 0.8478 \end{bmatrix} \tag{10.14}$$

As pointed out in Chapter 8, the importance of each block or data set can be measured by the score block weight as in Equation (10.8). Clearly, the score block weights for the synchronous generators are larger than for the wind farms for the first four components as expected from physical considerations.

Figure 10.14 depicts the shape associated with the dominant mode in the multiblock decomposition. At a glance, the result coincides with the mode shape in Figure 10.3.

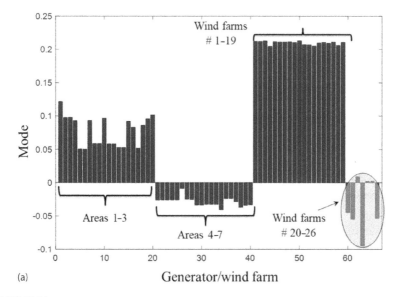

(a)

FIGURE 10.14
Comparison of spatial patterns extracted using multiblock PCA and DMD analysis.
(a) Multiblock PCA   (*Continued*)

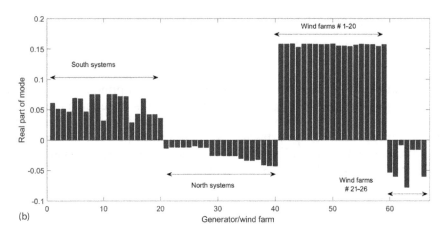

(b)

FIGURE 10.14 (Continued)
Comparison of spatial patterns extracted using multiblock PCA and DMD analysis. (b) DMD.

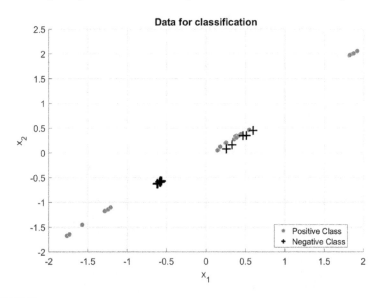

FIGURE 10.15
Scatterplot Pictorial representation of the test system showing approximate regional systems. Blue crosses correspond to positive values, red filled circles correspond to negative values.

Further insight into the coupled dynamics of wind machines and wind farms can be gained from the application of classification techniques. Support vector machine classification in Figure 10.15 shows that generator and wind farm dynamics are highly significant, at least for a group of them. These results correlate well with the shapes of the dominant eigenvectors in Figure 10.14, but the analysis allows the determination of machines and wind farms interacting.

References

Alonso, A. A., Frouzakis, C. E., Kevrekidis, I. G., Optimal sensor placement for state reconstruction of distributed process systems, *Process Systems Engineering*, 50(7), 1438–1452, 2004.

Arvizu, C. M. C., Messina, A. R., Dimensionality reduction in transient simulations: A diffusion maps approach, *IEEE Transactions on Power Delivery*, 31(5), 2379–2389, 2016.

Barocio, E., Pal, B. C., Thornhill, N. F., Messina, A. R., A dynamic mode decomposition framework for global power system oscillation analysis, *IEEE Transactions on Power Systems*, 30(6), 2902–2912, 2015.

Juarez, T., C., Messina, A. R., Ruiz-Vega, D., Analysis and control of the inter-area mode phenomenon using selective one-machine infinite bus dynamic equivalents, *Electric Power Systems Research*, 76, 180–193, 2006.

Messina, A. R., *Wide-Area Monitoring of Interconnected Power Systems*, IET Power and Energy Series 77, Stevenage, United Kingdom, 2015.

Messina, A. R., Vittal, V., Ruiz-Vega, D., Enríquez-Harper, G., Interpretation and visualization of wide-area PMU measurements using Hilbert analysis, *IEEE Transactions on Power Systems*, 21(4), 1763–1771, 2006.

Index